Optimization in Industry 2

Optimization in Industry 2

Mathematical Programming and Modeling Techniques in Practice

Edited by

Tito A. Ciriani
IBM Scientific & Technical Solutions, Rome, Italy

and

Robert C. Leachman
University of California at Berkeley, USA

JOHN WILEY & SONS
Chichester · New York · Brisbane · Toronto · Singapore

Baffins Lane, Chichester,
West Sussex PO19 1UD, England

National Chichester (0243) 779777
International (+44) 243 779777

Other Wiley Editorial Offices

John Wiley & Sons, Inc., 605 Third Avenue,
New York, NY 10158-0012, USA

Jacaranda Wiley Ltd., G.P.O. Box 859, Brisbane,
Queensland 4001, Australia

John Wiley & Sons (Canada) Ltd, 22 Worcester Road,
Rexdale, Ontario M9W 1L1, Canada

John Wiley & Sons (SEA) Pte Ltd, 37 Jalan Pemimpin #05-04,
Block B, Union Industrial Building, Singapore 2057

Library of Congress Cataloging-in-Publication Data

Optimization in industry 2 : mathematical programming and modeling techniques in practice / edited by Tito A. Ciriani and Robert C. Leachman.
p. cm.
"Proceedings of the 1992 IBM Institute on the subject of optimization"—
Includes bibliographical references and index.
ISBN 0 471 94192 1
1. Programming (Mathematics)—Congresses. 2. Mathematical models--Congresses. I. Ciriani, Tito A. II. Leachman, Robert C.
T57.7.0682 1994 93-5077
658.4′ 033—dc20 CIP

British Library Cataloguing in Publication Data

A catalogue record for this book is available from the British Library

ISBN 0 471 94192 1

Typeset in 10/12 Sabon by Dobbie Typesetting Ltd, Tavistock, Devon
Printed and bound in Great Britain by Bookcraft (Bath) Ltd

Contents

List of Contributors

Dr Alessandro Agnetis
Dipartimento di Informatica
e Sistemistica
Università di Roma 'La Sapienza'
Via Buonarroti 12
00185 Rome
Italy

Dr S. Ahn
Operations and Information Management
The Wharton School of the University
of Pennsylvania
1300 Steinberg Hall—Dietrich Hall
Philadelphia
PA 19104-6366
USA

Dr Ranga Anbil
American Airlines Decision Technologies
Dallas—Fort Worth Airport
Texas
USA

Dr Cynthia Barnhart
Department of Civil and Environmental
Engineering
Massachusetts Institute of Technology
Cambridge
MA 02139
USA

Dr Alessandro Ciancimino
Dipartimento di Informatica e
Sistemistica
Università di Roma 'La Sapienza'
Via Buonarroti 12
00185 Rome
Italy

Dr Tito A. Ciriani
Via Rigattieri 10
56127 Pisa
Italy

Dr Ronald Coxhead
Bonner & Moore Associates GmbH
Bannhoffstrasse 44/46
65185 Wiesbaden
Germany

Dr Christof Dillenberger
IBM German Manufacturing Technology
Center
Dept. 0817: Modeling Methods
Max-Eyth Strasse 6
71065 Sindelfingen
Germany

Dr Roland Ernst
Ford Motor Company
John-Andrews-Entwicklungszentrum
Dep. MD//PN-320
5000 Köln
Germany

Dr Laureano F. Escudero
UITESA, Iberdola group and
Universitad Complutense de Madrid
Cindad Universitaria
28040 Madrid
Spain

Dr Giorgio Fasano
Alenia Spazio S.p.A.
Corso Marche 41
10146 Turin
Italy

Dr E. H. Frazelle
School of Industrial and Systems Engineering
Georgia Institute of Technology
Atlanta
GA 30332
USA

Dr Stefano Gliozzi
IBM Semea
Centro Sviluppo Software Applicativo
Viale Oceano Pacifico 173
00144 Rome
Italy

Dr Monique Guignard-Speilberg
Wharton School
University of Pennsylvania
Philadelphia
USA

Dr Steven T. Hackman
School of Industrial and Systems Engineering
Georgia Institute of Technology
Atlanta
GA 30332
USA

Dr Levent Hatay
Kader Sokak 20/2
Gazi Osman Pasa
06700 Ankara
Turkey

Dr Hua Thanh Binh
Chairman of the HUA Group
21 Boulevard Delessert
75016 Paris
France

Prof. Ellis L. Johnson
IBM T. J. Watson Research Center
Mathematical Sciences Department
Yorktown Heights
New York 10598
USA

Prof. Robert C. Leachman
Engineering Systems Research Center
University of California
Berkeley
CA 94720
USA

Dr Mark Lembersky
Chairman of the Lembersky Chi Inc.
651 Strander Boulevard
Seattle
WA 98188
USA

Prof. Mario Lucertini
Dipartimento di Ingegneria Elettronica
II Università di Roma
Via della Ricerca Scientifica
00133 Rome
Italy

Dr J. Meyer
Ford Motor Company
John-Andrews-Entwicklungszentrum
Dep. MD/PN-320
5000 Köln
Germany

Prof. Paolo Migliavacca
Via Silvio Pellico 14
20050 Lesmo
Milan
Italy

Dr U. Passy
School of Industrial and Systems Engineering
Georgia Institute of Technology
Atlanta
GA 30332
USA

Dr Massimo Pizzichella
IBM Semea
Direzione Centrale Ricerca e Innovazione Tecnologica
Viale Oceano Pacifico 173
00144 Rome
Italy

Dr L. K. Platzman
School of Industrial and Systems Engineering
Georgia Institute of Technology
Atlanta
GA 30332
USA

Mr Dale J. Raar
Harris Corporation—Semiconductor Sector
Manufacturing Systems, MS 62A-024
2401 Palm Bay Road
Palm Bay
FL 32905
USA

Dr Anna Sciomachen
Dipartimento Scienza dell'Informazione
Università degli Studi di Milano
Via Comelico 39/41
20135 Milan
Italy

Prof. H. P. Williams
Faculty of Mathematical Studies
University of Southampton
Southampton
SO9 5NH
UK

Dr Arthur Wollensak
IBM German Manufacturing Technology Center
Dept 0817, Modeling Methods
Max-Eyth Strasse 6
7032 Sindelfingen
Germany

Preface

1992 marked the eleventh anniversary of the IBM Europe Institute, an annual forum highlighting current important topics in scientific and technical computing. The "Optimization Solutions" session of the IBM Europe Institute was held August 3–7, 1992 in Oberlech, Austria. This seminar included speakers from industry and academia in Europe as well as the United States. Its purpose was to provide an overview of the entire cycle of development, implementation and beneficial operation of mathematical optimization applications in industry. Speakers presented strategies and techniques embodied in useful mathematical programming applications in various industries, as well as recent advances in mathematical modeling and computer software with significant implications for industrial applications of optimization.

Thirteen of the presentations at the seminar and two additional contributions have been collected in this volume. A diverse set of application areas and technical modeling approaches are represented, reflecting the diversity of the participants and the wide scope of potential applications and problem structures for mathematical optimization.

As the ability to solve integer and non-linear optimization problems has improved, the applications of optimization to product design have been growing. Interactive systems that employ embedded optimization models for the design of home improvement structures and office layouts have been developed by Lembersky Chi. For example, interactive systems in use at home improvement centers in the United States feature combinatorial optimization models that generate structural designs for home improvements considering applicable building codes, sound engineering principles, ease of construction, and low cost.

In another application of non-linear programming to product design, Ford Motor Company in Germany has developed models to optimize valve train motion in automotive engines.

The scheduling of transportation operations is a growing area for profitable application of optimization. Researchers from Georgia Tech and IBM Research have worked with American Airlines Decision Technology to develop and implement large-scale integer programming models that assign one crew to each scheduled flight such that the overall costs are minimized and work rule restrictions are satisfied.

Optimization arises naturally in any context involving the allocation of scarce resources or funds among competing demands or suppliers. The University of Southampton has investigated the selection and allocation of contractors to public works projects in the United Kingdom. The problem, initially formulated as an integer programming model, was shown to be a network flow model readily solvable on a personal computer.

Another example of optimization applied to an allocation problem involves the selection of payloads for the Space Station Freedom. Alenia Spazio developed a large-scale integer programming application used to select which research experiments to install subject to constraints for space, power, heat rejection, microgravity and crew time.

The allocation of storage space in distribution systems is another subject for optimization. The Material Handling Research Center of Georgia Tech has developed a system for the allocation of stock-keeping units to forward and reserve picking zones and to various storage media so as to minimize costs. The system incorporates a non-linear integer programming model.

Production planning and scheduling is a traditional area for the application of optimization techniques. The planning and scheduling of fertilizer production at Enichem Agricoltura illustrates the planning complexities that arise when modelling chemical process and distribution problem. An overall methodology for planning and scheduling was described that interfaces optimization models for long-term and medium-term planning based on standards with a more detailed daily plant production.

For the refinery process Bonner & Moore developed an integrated planning and scheduling system where both planning and blend optimization benefit from advances in algorithms for the solution of the non-linear mixed integer programming problem.

Increasing importance of on-time delivery performance and control of lead times in manufacturing have provided strong incentives for the application of optimization techniques to production planning and delivery quotation. A planning system incorporating rate-based linear programming models of semiconductor production was developed by the University of California at Berkeley and implemented at Harris Corporation–Semiconductor Sector with impressive results.

Many manufacturing scheduling problems provide challenging integer programming formulations. For example, modelling the punching of multi-layer ceramics in electronics manufacturing leads to a formulation with numerous binary variables representing potential production setups and tool changes. An optimization model developed and applied at IBM Sindelfingen in Germany was described that starts with a linear programming solution and then successively forces the integrity of binary variables so as to obtain an efficient production schedule.

Another combinatorial scheduling problem arises in the assembly of electro-mechanical systems such as robots, where it is desired to sequence assembly tasks so as to minimize completion time. An effective heuristic approach to this problem has been developed by the University of Milan.

The design of assembly lines, in which production tasks are assigned to machining stations, is another type of optimization problem arising in assembly. Researchers at University of Rome have developed dynamic programming models for optimizing the design of assembly lines producing automotive components.

Inventory control is one of the oldest areas for the application of scientific management, but the recent advance of computers and information systems make possible the consideration of multi-company supply chains and the optimization of inventory controls within such chains for the benefit of all participants. A Decision Support System has been developed by Hua Group for order optimization at all levels of a multi-partner manufacturing, distribution and retail supply chain.

Financial investment is a traditional resource allocation problem for which researchers are developing new optimization-based techniques. A research team based at the Wharton School of Pennsylvania University and at Uitesa in Spain has proposed several mixed 0–1 programs for allocating funds among financial instruments with uncertainty in interest rates, analyzed using scenarios.

Finally, paralleling the expanding use of optimization in industry is a growing need for modeling software, i.e., user interfaces and user tools to facilitate the formulation and operation of optimization models in industrial practice. IBM Semea has developed an interactive language for algebraic formulation of optimization models.

Tito A. Ciriani
Robert C. Leachman

Disclaimer

In relation to the following proceedings of the 1992 IBM Europe Institute on the subject of optimization, IBM wishes to draw to the reader's attention the fact that the papers that appear in this book have been developed independently by the authors concerned and do not necessarily represent IBM's views on the subjects described.

1 Optimization at the Customer–Organization Interface

Mark Lembersky
Lembersky Chi Incorporated, Seattle, WA, USA

ABSTRACT
The development of user-friendly systems for office layout and for design of home improvements are described. These systems embed optimization models that strive to achieve an overall user-specified design, subject to constraints modeling limitations such as building codes, sound engineering principles and acceptable costs. Follow-on optimization calculations are employed to plan material procurements so as to minimize waste and cost.
These systems illustrate the challenges and opportunities for developing applications of optimization at the customer interface.

1 INTRODUCTION

As we know, all organizations are encountering major challenges in the 1990s: escalating competitive pressures, increasing customer expectations, and a poorly educated, shrinking workforce. This is equally true for industrial and service organizations and in both retail and business-to-business situations.

The challenges are especially great where satisfying the customer involves a complex product or service, tailored to each customer's needs. Two such areas that I have been involved in will be used for illustration: first, the design and retail sale of home-improvement projects for consumers looking to modify their houses and, secondly, the layout and selling of office furniture to businesses in the process of renovating their office areas or moving into new or expanded space.

Optimization in Industry 2. Edited by T. A. Ciriani and R. C. Leachman

While the challenges are great, so are the opportunities for utilizing optimization tools. And, the feasibility of using such tools is dramatically enhanced today because of the widespread introduction of computers 'in the front lines' of organizations, making it possible for the first time to gather the extensive data that most optimization models and algorithms require.

Still, *successful* examples of such optimization applications at the interface between the organization and its customer are rare. This is often because the introduction of computer-based optimization makes the situation worse, as employees, salespeople and customers are unable or unwilling to effectively utilize the technology.

Of course, it *is* possible to design and effectively implement computer-based optimization tools that people at those front lines, even math-phobic and computer-naive ones, master, utilize, and even enjoy—and that help the organization meet its objectives. To describe our approach to achieving these goals, I'll use two recently completed, real implementations; one in the home-center industry and the other in the office-furniture industry.

2 TWO APPLICATION AREAS

A little background: with the high and rising cost of housing, home improvement is a major activity for North American families, and is projected to grow beyond $100 billion a year. Building an outdoor deck is the most frequent do-it-yourself project in the United States. Other popular projects include adding a garage, installing large shelving units, erecting fences, and finishing basements.

There are well over 20 000 retail 'home-center' stores in North America that sell lumber, hardware and other building materials to do-it-yourselfers. With increasing frequency, in order to remain competitive, home centers must be able to help consumers with all aspects of a potential project, not just sell them the individual pieces. This means salespeople are called on to undertake complicated design problems that require knowledge of structure engineering and construction techniques, plus be able to accurately estimate all the materials needed and give potential customers confidence the project will turn out meeting their desires. Yet, almost all salespeople have limited educational backgrounds, experience, or training for these tasks. Further, home centers—like all organizations—are facing a shortage of motivated, long-term workers.

And, as in the popular Neil Diamond song, 'Change a few of the names and the story's the same' in retailing, business-to-business selling, on the factory floor, and in service organizations.

To illustrate in the business-to-business arena, consider office furniture, where annual sales in the United States are $7.5 billion. All such sales require someone to lay out a workable office plan in advance. More and more, businesses go to those 10% of the 20 000 furniture suppliers who can provide these design services. While the furniture suppliers' representatives have more training than their counterparts at home centers, the number of options to be considered is much greater, the clients are more sophisticated, and the design is a very complex three-dimensional spatial-arrangement problem, with many office linkages and qualitative factors that must be reflected in the final design. Further, time is at a premium for these very busy representatives. They rarely have the time to identify and evaluate all the possible options that might satisfy the customer's needs.

Increasingly popular modular furniture has even more possibilities. But modular office furniture fits together in certain ways and not in other ways. A design that looks great on paper will be a failure if it cannot be easily assembled. Further, having completed a conceptional design, the representative is left with the daunting task of determining all of the components needed—down to the last panel connector.

3 THE SYSTEMS

For more than 15 years, much of my attention has focused on developing customized, computer-based optimization systems that attempt to make it easy for people without special training to perform complex calculations and to visualize and explore paths of action. My current firm and its predecessor have had the opportunity to develop such systems for the situations I have described.

The system for home improvement was custom-created on behalf of Weyerhaeuser Company (an international wood-products supplier), and that for office layout was custom-developed for Phoenix Designs (a subsidiary of Herman Miller, a leading supplier and innovator of modular office furniture).

The algorithms and associated software in the home-improvement system allow salespersons and consumers to rapidly sketch home-improvement projects, make key feature decisions, and then literally 'see' the finished project in an on-screen, dynamic three-dimensional color representation. At the heart of the software are embedded optimization algorithms that find the best structural design. Additionally, other optimization algorithms calculate the materials required, trading off material-usage efficiency and cost with the ease of implementing cutting patterns. (See Lembersky, 1991, for illustrations of the system's screens.)

4 DEVELOPMENT

There were significant technical and human obstacles to overcome in developing and implementing these systems. In each case, algorithms were necessary that would take any overall design outlined and find a detailed design that was feasible. For example, for home improvement this meant a structural design that simultaneously meets each of the following requirements, expressed as constraints:

- meet national building codes;
- satisfy sound engineering principles, which can often be either stronger or weaker than code requirements;
- produce a structural design that is relatively easy for a do-it-yourself consumer to construct;
- have a competitive cost for the consumer's total project.

These constraints are in conflict. For example, the simpler the design is to build, the more material it will require (thus increasing cost) in order to meet code and engineering principles.

Generally available CAD (computer-aided design) software implicitly assumes an informed user; for example, it assumes the user knows building codes and engineering principles and will reflect this in any design. Also, CAD software assumes that whatever the user designs, the user is able and willing to pay for and install. In other words, traditional CAD algorithms are useful primarily to technically sophisticated and computer-comfortable designers and engineers. Clearly, they are not suitable for the home-improvement or the office layout situation.

Consequently, we found it necessary for each system to develop what some may call expert systems that develop a design meeting all relevant constraints. However, this embedded system asks no questions of the customer. The customer and salesperson outline their design choices and the algorithms return the appropriate complete design.

Once the design is complete, another problem is determining the complete list of materials needed. The solution becomes the list of items printed out for the customer to purchase.

Let's look at the home-improvement system. In North America, lumber comes in a small set of discrete lengths, generally 8 feet through 16 feet in 2-foot increments. Of course, the actual lengths needed in constructing a project will rarely match these available lengths. A typical project involves hundreds of needed lengths; thus, for example, simple rules such as rounding up each needed length to the first larger available length can lead to considerable waste and increased cost. More complex procedures are needed. Thus, we have several 'knapsack' optimization problems that must be solved. In addition, some materials needed for projects do not come from lumber, but rather from thin, 4-feet wide by 8-feet long sheets. The cutting pattern used to extract needed materials from such sheets must be efficient in order to control the cost of the project. Thus, we have a 'cutting-stock' optimization problem for each sheet. Complicating both the knapsack and cutting-stock problems is the need to have the final patterns be simple enough for a consumer to easily extract the materials needed. This may mean, for example, 'backing off' from an otherwise optimal use of material to patterns that produce a somewhat higher cost, but are unlikely to lead to consumer error or difficulty in execution.

Analogous algorithms were needed for the office application.

Having these optimization algorithms embedded in the software relieves the organization's employees of the need to have certain skills and capabilities, yet keeps them 'in the loop' and in charge of the interaction with the customer. It also illustrates the principle:

- assign to technology what technology does well: fast, accurate arithmetic, mindless instruction following—which, serendipitously, is what people do not do well or like to do; and
- reserve the rest for humans, especially guidance of the process—which humans enjoy and, in my opinion, will continue to do better than computers.

I feel that this partition of responsibility between humans and technology is necessary for *success*, even if it means humans make what algorithmically are suboptimal decisions.

In the home-improvement system, a deck can be rotated smoothly in any direction, so that the consumer may first look underneath to see the support structure and then decide to see how the project will look from the neighbor's backyard. An office can be revolved to see how it will look from the front door, from a boss's office, or from any other angle. This capability was not included to demonstrate our ability to get very high graphical performance from economical technology, but because it was needed in order for the system to be effective. Einstein said: 'Everything should be as simple as possible, but no simpler.' The same is true, I believe, for successfully applying optimization: incorporate all the methodology that is needed, but only that which is truly required.

5 IMPLEMENTATION

A fact of life we all need to deal with is that many people whom applications target are turned off by mathematics and are neither computer-literate nor interested in becoming so. In this day and age, it is increasingly apparent that many people feel threatened or overwhelmed by the technology around them, with no personal way to take advantage of it. Consequently, achieving success means not only solving all the technical problems involved, but doing so in a way that deals effectively with these 'technology issues'. A system must not just do what it was designed to do, but must do so in a way that is attractive and enjoyable to employees and, when relevant, customers—resulting in a positive experience with the technology and, therefore the organization.

As a result, considerable attention was focused on making systems as easy to use as a video game. Previously, decision simulators (Lembersky and Chi, 1984, 1986) and visual interactive modeling (surveyed by Bell, 1985) had been used in related situations in manufacturing. We applied these ideas to retailing and business-to-business selling, apparently for the first time. This resulted in a unique 'look and feel', which could not be provided by the conventional graphics tool kits. For example, there are no 'user's manuals' for operating the systems. Yet, for over 200 installations, people have become proficient in 15 minutes or less. People report finding such systems inviting to use, even if they are reluctant to try other computer-based systems. The computer becomes invisible and the optimization computing is made natural.

6 IMPACTS

A few words on impacts, which I can discuss only for the home-improvement system. In summary terms, by end of the first season, $150 million of projects had been designed. Comprehensive surveys of retailers, salespersons, and consumers revealed a 270% increase in the 'close rate' and almost 250% the number of home-improvement project designs than at comparable stores without the system. These same surveys report that this interactive optimization system made salespeople at least 40% more productive. Employees themselves reflected improved skills, professionalism, self-image, and job satisfaction. Survey data also shows increased customer loyalty and store image. Nine out of every ten consumers said they were more likely to return to a store after using the system there. And 100% of interviewed consumers said they would recommend the system to a friend.

7 CONCLUSION

I'll close with the comment that beyond these systems, the combination of optimization and interactive computing developed for these applications can be beneficial for worker skill enhancement, productivity increase, and job enrichment in many organizational settings.

REFERENCES

Bell, P. C. (1985) Visual interactive modeling as an operations research technique, *Interfaces*, **15**, 26–33.

Lembersky, M. R. (1991) Enhancing point-of-sale productivity and satisfaction in retail stores, *European Journal of Operations Research*, **54**, 325–329.

Lembersky, M. R., and Chi, U. H. (1984) Decision simulators speed implementation and improve operations, *Interfaces*, **14**, 1–15.

Lembersky, M. R., and Chi, U. H. (1986) Weyerhaeuser decision simulator improves timber profits, *Interfaces*, **16**, 6–15.

Lembersky Chi Incorporated
651 Strander Boulevard
Seattle
WA 98188
USA

2
A Column-Generation Technique for the Long-Haul Crew-Assignment Problem

Cynthia Barnhart[1], Ellis L. Johnson[2], Ranga Anbil[3], and Levent Hatay[3]
[1]*Massachusetts Institute of Technology, Cambridge, MA, USA*
[2]*IBM Watson Research Center, New York, and Georgia Institute of Technology, Atlanta, GA, USA*
[3]*American Airlines Decision Technologies, Dallas, TX, USA*

ABSTRACT

The crew-assignment problem is characterized by a set of flight segments, work rules restricting crew utilization, crew-availability limitations, and costs for every flight expressed as a function of time. The objective is to assign one crew to each scheduled flight such that overall costs are minimized and crew-availability and work-rule restrictions are satisfied. A long-haul crew-assignment problem is characterized by flight segments that are typically long-distance and do not operate on a daily schedule.

The long-haul crew-assignment problem is solved by a two-phase process. In the first phase, the integrality constraints are relaxed and the resulting linear program is solved using column generation. Columns are efficiently determined using a specialized shortest-path procedure on a long-haul network. The network is constructed so that certain network paths correspond to crew pairings. In the second phase of the solution procedure, an integer crew assignment is determined. Alternative techniques for achieving these solutions are discussed. Finally, computational experience obtained in solving the crew-assignment problem of a long-haul operating airline is reported.

1 INTRODUCTION

The inputs to the long-haul crew-assignment problem include a planning period; *crewbase* locations; and flight services, typically international, with scheduled departure and arrival times at specified origin and destination points. The objective

Optimization in Industry 2. Edited by T. A. Ciriani and R. C. Leachman

is to assign one crew to cover each flight such that overall costs are minimized and crew-availability and work-rule restrictions are satisfied.

To achieve this, crews are assigned to *pairings*, where a pairing is a sequence of flights with the first flight beginning and the last flight ending at the same crewbase. The sequences of flights are grouped to form *duty periods* separated by rest periods called *overnights*. The rules restricting duty and rest periods are described in detail in the next section. The total duration of a pairing, that is, elapsed time from departure to return of the crew, is called the *time-away-from-base* (TAFB). TAFB for each pairing cannot exceed some maximum number of days as mandated by work rules.

We develop a price-directive column-generation approach to solve the linear relaxation of the long-haul crew-assignment problem. Columns are generated using a specialized shortest-path procedure exploiting the time-based structure of the *long-haul network*. The network is constructed so that each *path* beginning and ending at the same crewbase with total length less than the maximum TAFB is a pairing. After determining a solution to the linear program (LP), an integer crew assignment is determined.

Confirmation of the concept of the solution procedure is provided by computational experience obtained in solving the crew-assignment problem of a long-haul operating airline.

In Section 2, rules governing the construction of pairings and their associated costs are described in detail. In Section 3, alternative representations of the long-haul network are presented. Solution procedures previously applied are reviewed in Section 4. Our problem formulation and the column-generation solution technique for the linear relaxation are detailed in Sections 5 and 6, respectively. Methods for achieving integer solutions are discussed in Section 7 and computational results are reported in Section 8. Finally, future research directions and conclusions are presented in Section 9.

2 CREW PAIRINGS

To solve the crew-assignment problem, a set of pairings must be constructed allowing an assignment of crews to pairings such that total costs are minimized, every flight is covered by one crew, and every crew is assigned to at most one pairing at any point in time. In this section, the work rules governing the definition of valid duty and rest periods are described and the costs to be minimized are defined.

Duty periods

Each duty period must satisfy work rules limiting the maximum number of flights in a duty period, the maximum flying time per duty period (as a function of the number of flights in a duty period), the maximum duration of a duty period, and the minimum and maximum connect time between consecutive flights in a duty period.

The flights making up a duty period can be classified into three types: operational, carryover, and deadhead. The operational flights are the flights scheduled during the planning period that must be assigned a crew.

Since crew pairings beginning during the current planning period may continue into the next planning period before returning to the crewbase, the *planning horizon* of concern extends to the maximum TAFB after the end of the current planning period. Flights scheduled in the follow-on planning period and included in the planning horizon are referred to as *carryover* flights. Carryovers may be, but need not be, covered.

A *deadhead* flight can occur in either of two ways: assignment of two or more crews to the same operational flight or assignment of a crew to a flight operated by another airline. The purpose of a deadhead flight is to reposition crews to increase their utilization.

Overnights

Work rules specify rest periods to be greater than some minimum time and less than a maximum time. The minimum rest time is a function of the duration of the duty period preceding the rest period, while the maximum time is station-specific. For duty periods with duration and flying time below some cutoff, a *short overnight* is required while duty-period durations exceeding this cutoff must be followed by a *long overnight*.

Costs

Pairing cost is usually based on flying time and/or TAFB as specified by union rules. The nature of the long-haul problem is such that the TAFB cost component is the dominant factor in determining overall cost. Therefore, the cost associated with each flight or overnight in a long-haul operation is expressed as a linear function of time and the objective is to minimize total TAFB.

Included in our model are additional costs associated with deadheads and credits associated with carryovers. A crew deadheading on its own aircraft may incur no additional cost, whereas a crew deadheading on a competitor's aircraft is charged the ticket price for that flight. In contrast, a crew assigned to a carryover flight receives a credit. If no credit were given, pairings would be terminated as quickly after the end of the planning period as possible. This strategy, while minimizing the current planning period costs, would result in higher crew costs in the following planning period. To ensure a more global approach, credit is given for covering carryovers. This credit, however, is limited to only one crew.

In summary, the objective is to determine a set of pairings covering each operational flight and minimizing total TAFB costs plus deadhead costs less carryover credits.

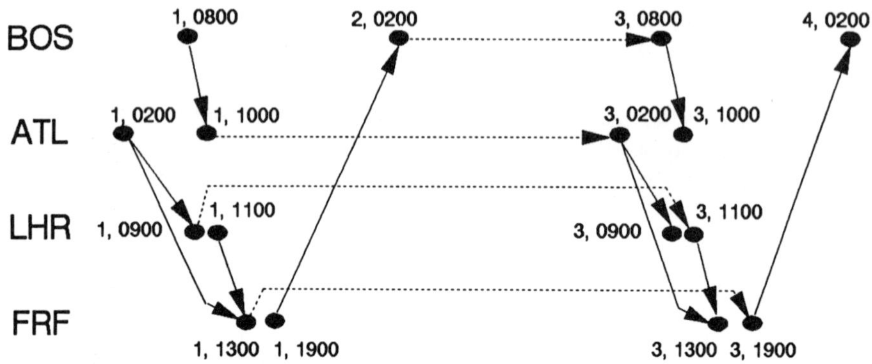

Scheduled Flights (GMT)

		Departure		Arrival				Departure		Arrival	
Orig.	Dest.	Day	Time	Day	Time	Orig.	Dest.	Day	Time	Day	Time
BOS	ATL	1	0800	1	1000	FRF	BOS	1	1900	2	0200
BOS	ATL	3	0800	3	1000	FRF	BOS	3	1900	4	0200
ATL	LHR	1	0200	1	0900	LHR	FRF	1	1100	1	1300
ATL	LHR	3	0200	3	0900	LHR	FRF	3	1100	3	1300

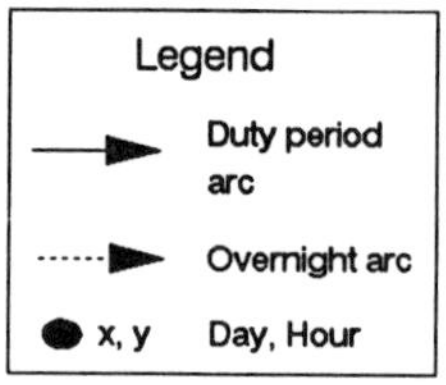

Figure 2.1

3 NETWORK REPRESENTATIONS

We develop two alternative network representations, the time-space and a time-line network for the long-haul crew assignment problem. The networks are detailed in the following sections.

3.1 Time-space network

The network depicted in Figure 2.1 contains two types of arcs: *duty-period* and *overnight* arcs. Given the set of flights scheduled during the planning period, every possible duty period is constructed and represented as one network arc. The tail node of each duty period arc is associated with the departure station and time of the first flight in the duty period and its head node is similarly associated with the arrival station and time of the last flight in the duty period.

Connecting duty-period arcs are the overnight arcs modeling rest time. For any two duty periods *a* and *b*, an overnight arc from the head node of *a* to the tail node of *b* is possible only if the arrival station of *a* is the departure station of *b* and the time between the departure of *b* and the arrival of *a* is greater than the minimum required rest time and less than the maximum rest time allowed at the station.

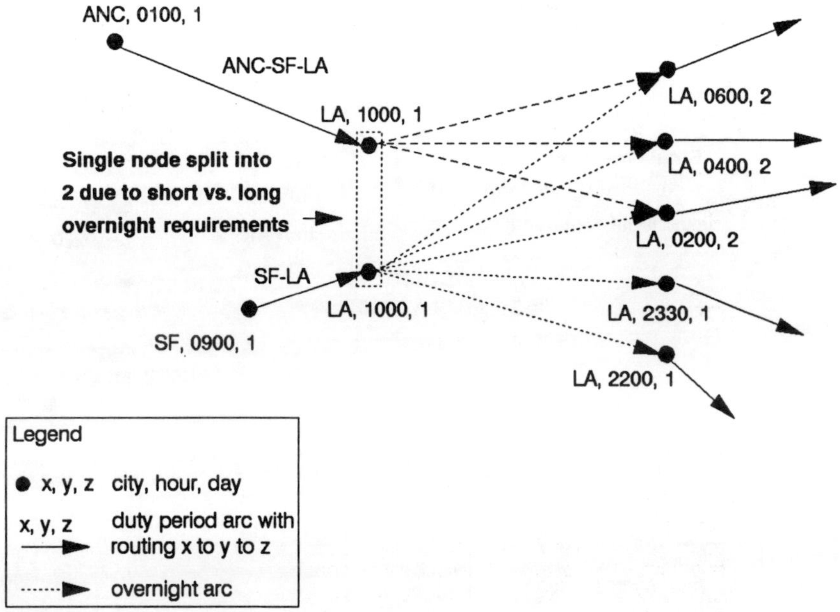

Figure 2.2

Importantly, each duty period beginning with a common flight has the same tail node. Similarly, except under one condition, each duty period ending with a common flight has the same head node. The exception occurs when duty periods ending with a particular flight vary in duration such that at least one of the duty periods requires a long overnight while the remaining duty periods need only a short overnight. In this case, as shown in Figure 2.2, the single head node is split into two nodes, a long-overnight and a short-overnight node. All *long duty periods* terminate with the long-overnight node and all others terminate with the short-overnight node.

Observe that each network path beginning and ending at the same crewbase with length less than maximum TAFB is a pairing. This special network construction facilitates the application of an optimal column generation approach to solve the LP relaxation of the long-haul crew assignment problem.

Arc costs and bounds

Each arc in the network has assigned cost equal to the difference between the times associated with its head and tail nodes multiplied by a factor F to convert the time measure into dollars.

The lower bound on flow for all arcs corresponding to deadhead and carryover flights is 0. However, the lower bounds for arcs corresponding to operational flights cannot be applied to a single arc but rather must apply to a set of *bundled arcs*. That is, the total flow summed over all duty-period arcs including any particular

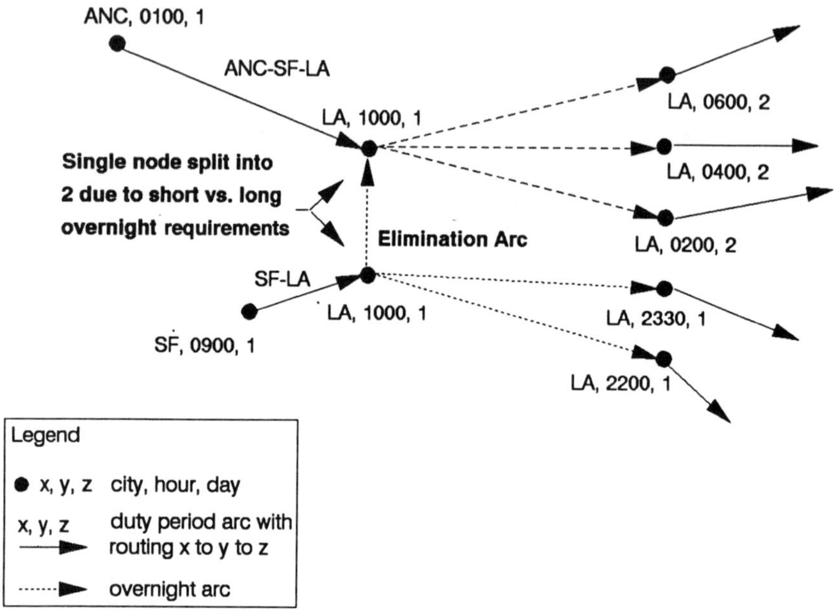

Figure 2.3

operational flight is 1. Similarly, for upper bounds, the maximum flow over all duty-period arcs including any particular operational, carryover, or deadhead flight cannot exceed the capacity (measured in crews) of the aircraft type covering the flight.

The lower bound for all overnight arcs is zero and the upper bound is unrestricted.

Network size

The time-space network contains one arc for each duty period and one arc for each legal rest period separating two duty periods. For long-haul problems, flights are long relative to maximum duty-period flying time and duration and thus, the number of flights that can be paired to form a duty period is limited. As a result, an explosion in the number of duty-period arcs does not occur.

The number of overnight arcs, however, grows rapidly as the number of flights in the long-haul problem increases. A method of reducing the size of the time-space network involves exploiting the long-overnight/short-overnight node relationships. Observe that each feasible rest period following a long duty period ending with some flight a is feasible for a short duty period ending with the same flight a. In the illustration in Figure 2.2, each rest period that can be preceded by either a long or a short duty period is represented as two arcs: one from the short-overnight node and one from the long-overnight node. Instead, as shown in Figure 2.3, the set of arcs corresponding to long overnights beginning

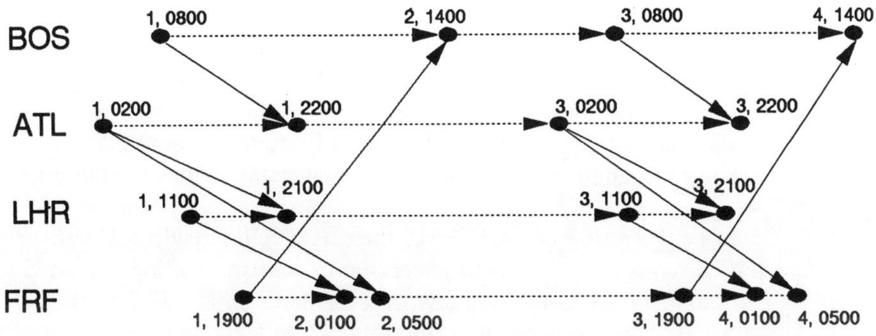

Scheduled Flights (GMT)											
		Departure		Arrival				Departure		Arrival	
Orig.	Dest.	Day	Time	Day	Time	Orig.	Dest.	Day	Time	Day	Time
BOS	ATL	1	0800	1	1000	FRF	BOS	1	1900	2	0200
BOS	ATL	3	0800	3	1000	FRF	BOS	3	1900	4	0200
ATL	LHR	1	0200	1	0900	LHR	FRF	1	1100	1	1300
ATL	LHR	3	0200	3	0900	LHR	FRF	3	1100	3	1300

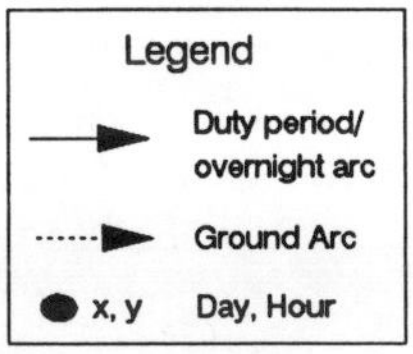

Figure 2.4

at a short-overnight node can be replaced with one arc from the short- to the long-overnight node.

3.2 Time-line network variant

Even with arc elimination, the number of overnight arcs in the time-space network may become unmanageable with increases in the number of flights. Thus, for problems containing a large number of flights, we propose the alternative network structure depicted in Figure 2.4. The network contains two types of arcs: duty-period/overnight arcs and ground arcs. As in the time-space network, a network arc with end nodes representing a station at a point in time is included for each duty period. The tail node of a duty-period arc has associated time and location equal to the departure time and station of the first flight in the duty period. Similarly, the head node of a duty-period arc has associated location equal to the arrival station of the last flight in the duty period. However, unlike in the time-space network, the time associated with the head node is not the arrival time of the last flight in the duty period but rather, it is the arrival time *plus* the minimum required rest, or overnight, time.

As in the time-space network, all duty periods beginning with the same flight have a common tail node; all duty periods ending with the same flight and requiring only a short overnight have a common head node; and all duty periods ending with the same flight and requiring a long overnight have a common head node.

To connect the duty-period/overnight arcs, all nodes (whether corresponding to an arrival or a departure) are sorted in increasing order by station and a ground arc is placed between adjacent nodes at the same station. This time-line at each station allows a crew upon its arrival into the station to depart at any point in the future. Thus, the time-line network representation, unlike the time-space network, does not guarantee satisfaction of the rule limiting the duration of the maximum rest period. However, the objective of minimizing total TAFB will work to eliminate from the solution the illegal pairings containing extended rest periods.

As in the time-space network, there is a relationship between network paths and crew pairings. With the exception of paths including extended rest periods, each path in the time-line network beginning and ending at the same crewbase with length not greater than maximum TAFB is a pairing.

Arc costs and bounds

Costs and bounds for the duty-period/overnight arcs in the time-line network are defined exactly as those for the duty-period arcs in the time-space network. Ground arcs in the time-line network are treated like overnight arcs in the time-space network and therefore, have lower bounds of 0 and unrestricted upper bounds.

Network size

The time-line network contains one arc for each duty period and one ground arc for each node. Thus, while the number of duty-period arcs in the time-space network equals the number of duty-period/overnight arcs in the time-line network, the number of ground arcs in the time-line network is far less, for most applications, than the number of overnight arcs in the time-space network. Additionally, the number of ground arcs in the time-line network can be reduced through node consolidation (Figure 2.5). That is, consecutive arrival or consecutive departure nodes on the time-line at a station can be merged into one node and all ground arcs connecting the affected nodes can be eliminated.

4. HISTORICAL PERSPECTIVE

Substantial effort has been expended in developing solution techniques for crew-pairing problems (Anbil *et al.*, 1991b; Arabeyre *et al.*, 1969; Baker and Fisher, 1981; Baker *et al.*, 1979; Ball and Roberts, 1985; Barutt and Hull, 1990; Crainic and Rousseau, 1987; Desrochers and Soumis, 1989; Etschmaier and Mathaisel, 1985; Gershkoff, 1989; Marsten and Shepardson, 1981; Minoux, 1984; Odier *et al.*, 1983; Rannou, 1986; Rubin, 1973; Shepardson and Marsten, 1980; and Thiriez, 1969). In the early 1970s, American Airlines used a column-generation solution strategy, called TRIP (described in Anbil *et al.*, 1991a), to heuristically

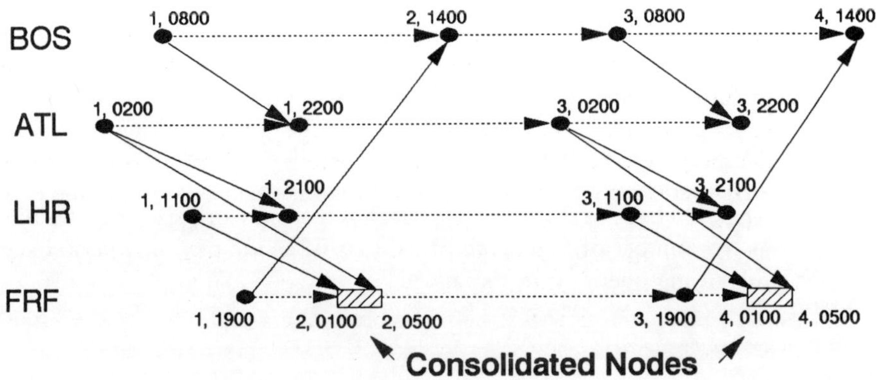

Scheduled Flights (GMT)

Orig.	Dest.	Departure Day	Departure Time	Arrival Day	Arrival Time	Orig.	Dest.	Departure Day	Departure Time	Arrival Day	Arrival Time
BOS	ATL	1	0800	1	1000	FRF	BOS	1	1900	2	0200
BOS	ATL	3	0800	3	1000	FRF	BOS	3	1900	4	0200
ATL	LHR	1	0200	1	0900	LHR	FRF	1	1100	1	1300
ATL	LHR	3	0200	3	0900	LHR	FRF	3	1100	3	1300

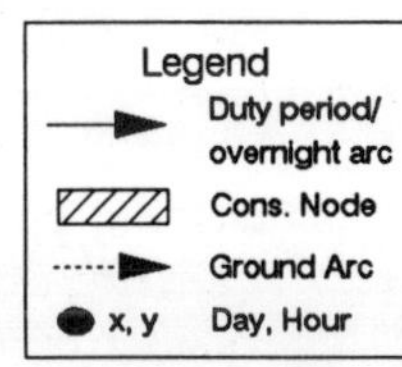

Figure 2.5

determine a solution for the daily domestic crew-scheduling problem. More recently, Lavoie *et al.* (1988) and Desrosiers *et al.* (1991) have described column-generation procedures that determine optimal solutions to the LP relaxation of the crew-scheduling problem. LP columns correspond to pairings and pairings are generated using constrained shortest-path procedures (such as those described in Desrochers and Soumis, 1989) over a network containing flight, connection, and overnight arcs. The shortest-path procedures are tailored to determine paths satisfying maximum duty-period flying time and maximum pairing-length requirements. Results for medium and long-haul problems with schedules of one and two weeks are presented in Desrosiers *et al.* (1991).

Earlier work by Minoux (1984) similarly involves the formulation of the crew-assignment problem in which each column represents a pairing. The solution procedure generates columns using a shortest-path procedure over a network where the arcs correspond to duty periods and the paths correspond to crew pairings. However, to model the rules defining pairings, Minoux's network includes several copies of each duty-period arc and several copies of the associated end nodes.

Rannou (1986), building upon Minoux's work, represents each possible duty period as a node, not an arc, and represents contractual links between duty periods as arcs. By replicating the nodes and arcs of the network as in Minoux (1984), the set of valid crew pairings is the set of all paths starting and ending at the same crewbase with length less than the maximum duration of a crew pairing. The practical consequence of the node and arc replication is that the network can become large. Rannou's computational experiments on the long-haul problem show the following relationship between number of flights and network size:

Number of flights	Network size: Number of nodes	Number of arcs
142	1333	11452
316	1798	47356
318	2138	31253

The drawback to Rannou's approach is that network size becomes unmanageable as the number of flights increases, that is, as the planning period is extended or the airline operation is expanded.

5 FORMULATION

The long-haul crew-assignment problem can be formulated as follows:

$$\text{LHCA: Minimize } \Sigma_{p\in p}\ \text{PC}_p X_p - \Sigma_{f\in C}\ \text{CR}_f c_f + \Sigma_{f\in D}\ \text{PP}_f d_f + \Sigma_{f\in O}\ M q_f \quad (1)$$

subject to:

$$\Sigma_{p\in p}\ X_p \delta_f^p + q_f \geqslant 1, \quad \forall f\in O \quad :\pi_f^O \quad (2)$$

$$\Sigma_{p\in p}\ X_p \delta_f^p \leqslant U_f, \quad \forall f\in O\cup C\cup D \quad :-n_f \quad (3)$$

$$\Sigma_{p\in p}\ X_p \delta_f^p - c_f \geqslant 0, \quad \forall f\in C \quad :\pi_f^C \quad (4)$$

$$\Sigma_{p\in p}\ X_p \delta_f^p - d_f = 0, \quad \forall f\in D \quad :\pi_f^D \quad (5)$$

$$\Sigma_{p\in p}\ \text{OFT}_p X_p \gamma_b^p \geqslant \text{MFT}_b, \quad \forall b\in B \quad :\propto_b \quad (6)$$

$$c_f \leqslant 1, \quad \forall f\in C \quad :-\lambda_f \quad (7)$$

$$X_p \text{ binary}, \quad \forall p\in P \quad (8)$$

$$c_f \geqslant 0, \quad \forall f\in C \quad (9)$$

$$q_f \geqslant 0, \quad \forall f\in O \quad (10)$$

$$d_f \geqslant 0, \quad \forall f\in D \quad (11)$$

where:

G: underlying graph with node set V and arc set A;
O: set of operational flights;
C: set of carryover flights;
D: set of deadhead flights;
F: set of flights, i.e. $F = O\cup C\cup D$;

B: set of crewbase locations;
P: set of all pairings;
ζ_f^a: $=\begin{cases} 1 \text{ if flight } f \text{ is included in arc } a, \text{ and} \\ 0 \text{ otherwise, for all } f \in F, \text{ for all } a \in A; \end{cases}$
β_a^p: $=\begin{cases} 1 \text{ if arc } a \text{ is included in path } p, \text{ and} \\ 0 \text{ otherwise, for all } a \in A, \text{ for all } p \in P; \end{cases}$
δ_f^p: $=\begin{cases} 1 \text{ if flight } f \text{ is included in path } p, \text{ and} \\ 0 \text{ otherwise, for all } f \in F, \text{ for all } p \in P; \end{cases}$
i.e., $= \Sigma_{a \in A} (\zeta_f^a)\beta_a^p$ since each flight is contained in at most one arc in any path;
γ_b^P: $=\begin{cases} 1 \text{ if pairing } p \text{ originates at crewbase } b, \\ 0 \text{ otherwise, for all } p \in P, \text{ for all } b \in B; \end{cases}$
U_f: upper bound on the number of crews assigned to operational, carryover, or deadhead flight $f \in F$;
CF: factor converting time-away-from-base to dollars;
FT_f: flying time for operational flight $f \in O$;
OFT_p: total operational flying time in pairing p,

$$\text{OFT}_p = \Sigma_{f \in O} \text{FT}_f \delta_f^p;$$

MFT_b: minimum total flying time required on pairings originating at crewbase $b \in B$ during the planning period;
T_a: elapsed time, i.e. head-node time minus tail-node time, for arc $a \in A$;
PC_p: pairing cost, i.e.

$$\text{PC}_p = (\text{CF})\Sigma_{a \in A} T_a \beta_a^p;$$

CR_f: credit for covering carryover flight f;
PP_f: purchase price of ticket on deadhead flight f;
M: large penalty cost, at least as great as CF multiplied by maximum TAFB;

with decision variables:

X_p: number of crews assigned to pairing p, for all $p \in P$;
c_f: number of crews (at most 1) that receive credit for covering carryover flight f, for all $f \in C$;
d_f: number of crews assigned to deadhead flight f, for all $f \in D$;
q_f: artificial variable ensuring coverage of operational flight f, for all $f \in O$;
π_f^O: dual variable associated with required coverage of operational flight f (constraints 2), for all $f \in O$;
n_f: negative of the dual variable associated with the restriction on the number of crews assigned to an aircraft (constraints 3), for all $f \in O \cup C \cup D$;
π_f^c: dual variable associated with coverage of carryover flight f (constraints 4), for all $f \in C$;
π_f^D: dual variable associated with coverage of deadhead flight f (constraints 5), for all $f \in D$; and
α_b: dual variables associated with the minimum requirements on total flying time at crewbase b (constraints 6), for all $b \in B$;
λ_f: negative of the dual variables associated with the restriction that credit for coverage of carryover flight f can occur at most once (constraints 7), for all $f \in C$;

Constraints (2) require that each operational flight be covered at least once. Over-coverage of operational flights, that is, more than one crew assignment to a flight, is permitted and represents the deadheading of crews on the company's aircraft. The number of crews assigned to a flight, however, is restricted to the capacity of the aircraft (constraints (3)). Constraints (4) state that carryover flights need not be covered. While coverage of the carryover flight is rewarded, over-coverage (constraints (7)) is not. Constraints (5) ensure that each deadhead is charged the ticket price for that flight. (The ticket price can be set to zero for deadhead flights on the company's own aircraft.) Crew-availability restrictions are modeled by constraints (6) requiring that the total flying time assigned to crews from each crewbase exceed some minimum amount during the planning period. This constraint ensures that the workload is spread equitably among the crewbases. Constraints (8) are the integrality constraints limiting the number of crews assigned to each pairing to 0 or 1. Finally, constraints (9), (10) and (11) are the nonnegativity constraints. The objective (1) is to minimize total cost as described in Section 2.

The long-haul crew-assignment problem LHCA is solved by a two-phase process. In the first phase, the integrality constraints are relaxed and the resulting linear program is solved using column generation. In the second phase, an integer solution is determined.

6 LP SOLUTION TECHNIQUE

The number of potential pairings for typical long-haul airline operations is likely measured in billions. Hence, it is neither attractive nor possible to solve problem LHCA directly. Instead, it is solved using a column-generation technique, specifically Dantzig–Wolfe decomposition (Dantzig and Wolfe, 1960). The master problem LHCA^i at iteration i is to solve LHCA using only a subset of all pairings. The subproblems are then to generate additional pairings for inclusion as columns in LHCA^{i+1}, the master problem at the next iteration. Termination occurs with an optimal solution to LHCA when the solution to LHCA^i at some iteration i satisfies the optimality conditions for all pairings, i.e.

$$\Sigma_{f\in O}\,(\pi_f^{O}-n_f)\delta_f^p+\Sigma_{f\in C}\,(\pi_f^{C}-n_f)\delta_f^p+\Sigma_{f\in D}\,(\pi_f^{D}-n_f)\delta_f^p+$$
$$\Sigma_{b\in B}\,\propto_b(\text{OFT}_p\gamma_b^p)\leqslant \text{PC}_p,\qquad \forall p\in P.$$

Rewriting:

$$l_p=\Sigma_{a\in A}\,\beta_a^P\,((\text{CF})T_a+\Sigma_{f\in O}\,\zeta_f^a(-\pi_f^O+n_f-\text{FT}_f(\Sigma_{b\in B}\,\alpha_b\gamma_b^P))+$$
$$\Sigma_{f\in C}\,\zeta_f^a(-\pi_f^C+n_f)+\Sigma_{f\in D}\,\zeta_f^a(-\pi_f^D+n_f))\geqslant 0\qquad \forall p\in P.$$

Thus, the subproblems $S(k)^i$ at iteration i for each tail node k located at a crewbase can be stated as follows:

$$S(k)^i\colon\ l^*(k)^i=\text{Min}_{p\in \text{PB}(k)}\ l_p,\qquad \forall k\in \text{CB},$$

where

CB: subset of V associated with any crewbase location $b \in B$; and
PB (k): set of all pairings beginning at node $k \in V$.

The subproblems can be efficiently solved with a shortest-path procedure for each node $k \in \mathrm{CB}$ using modified arc costs T'_a as follows:

$$T'_a = (\mathrm{CF})T_a = \Sigma_{f \in O}\ \zeta^a_f(-\pi^O_f + n_f - \mathrm{FT}_f(\Sigma_{b \in B}\ \alpha_b \gamma^p_b)) + \Sigma_{f \in C}\ \zeta^a_f(-\pi^C_f + n_f) + \Sigma_{f \in D}\ \zeta^a_f(-\pi^D_f + n_f)), \qquad \forall a \in A.$$

For operational flights, the modified costs computed for each shortest-path procedure are dependent on the crewbase from which the paths originate. That is, modified-arcs costs depend on the starting node of the shortest-path procedure.

If, for all $k \in \mathrm{CB}$, $l^*(k)^i \geqslant 0$, the crew-assignment problem LCHA is solved and the procedure terminates; otherwise, for all $k \in \mathrm{CB}$, a pairing p with $l_p = l^*(k)^i$ is added to the master problem LHCA^{i+1} and the entire process is repeated. The new master problem and the associated subproblems are solved and the process repeats until optimality is achieved.

In the next section, efficiency issues associated with solution of the subproblems are addressed.

6.1 Subproblem solution

The time-based network structure of the crew-assignment problem can be exploited to more efficiently solve the shortest-path subproblems. In particular, since no directed cycles exist in G, a topological ordering of the nodes in the long-haul networks can be achieved. Nodes in the network are sorted and numbered in order of increasing time. Then, a simplified Dijkstra (1959) shortest-path algorithm is applied. Specifically, the bottleneck operation of selecting the next node to scan is replaced with a simpler strategy. Rather than locating the unscanned node with the minimum-path label, the node-selection process simplifies to picking nodes in increasing order (Ahuja *et al.*, 1993).

Additionally, the simplified shortest-path procedure is modified to generate pairings exclusively. That is, the only paths considered are those beginning and ending at the same crewbase location with total duration less than the maximum allowable TAFB. The necessary modification is extremely easy to implement. To explain, let s denote the number of the node corresponding to the crewbase location where pairings are to originate, i.e. the root node of the shortest-path tree; and let t represent the largest node number with associated time less than or equal to that of s increased by the maximum TAFB. Then, the shortest-path procedure begins by selecting and processing nodes in ascending order beginning with node s and ending with node t.

The shortest-path procedure can also easily incorporate the requirements that a pairing begin and end at the same crewbase and not include an overnight rest at that crewbase. This is accomplished by not scanning the arcs out of any selected node with location equal to that of the origin of the pairing.

The resulting shortest-path procedure requires that each arc be examined at most once. Thus, the worst-case performance of the modified algorithm is O(A).

The reduction in run time associated with the simplified procedure can be significant since one shortest-path procedure for each tail node at a crewbase is run before solving problem $LHCA^i$, for every iteration i. This provides justification for expending effort in the network-construction stage so that shortest-path generation is as fast as possible. This differs from the approach used by Desrosiers *et al.* (1991), wherein the shortest-path procedure is modified to locate paths satisfying maximum duty time. Although this results in a more involved, time-consuming column-generation process, it does allow flexibility in the rules defining legal duty periods and pairings.

6.2 Initial solution

To obtain an initial solution for LHCA, the challenge is to generate *good* pairings to be used in the initial LP. There are many characterizations for good pairings. First, since the objective of minimizing total TAFB is similar to minimizing the total rest time in a pairing, good pairings have short rest periods. For similar reasons, good pairings have short connection times between flights and therefore, may include several *throughs*, i.e. two consecutive flights flown by the same plane. To generate pairings with these qualities, a time-space network is constructed, as described in Section 3.1, with the following restrictions:

(a) the only duty-period arcs included are those satisfying a user-constructed rule set specifying required and/or disallowed flight connections at a station; and
(b) the only overnight arcs included are those satisfying maximum rest-period allowances, as modified by the user for each station.

In the *restricted* time-space network *all* paths corresponding to pairings with restricted total TAFB, as defined by the user, are generated. These pairings are then included as the initial set of LP columns. Recall that feasibility is not an issue because artificial variables are included for each operational flight.

7 INTEGER PROGRAM SOLUTION

Given the LP solution, we use a general-purpose mixed-integer programming package to determine an integer solution to the long-haul crew-assignment problem. Our future research efforts will involve the development of a specialized procedure to determine integer solutions using column generation. The branching strategy, patterned after that of Ryan and Falkner (1988) and Anbil *et al.* (1991b) is to *fix a follow-on*, i.e. to either force the same crew to consecutively cover two flights or to force different crews to cover the two flights.

Since all pairings violating fixed follow-ons must be eliminated from the solution, new *legal* pairings will be generated using the specialized O(A) shortest-path procedure on an *altered* long-haul network. The network will be reconstructed so that any path corresponding to a pairing satisfies the fixed follow-ons. Deadhead flights on the company's aircraft add to the complexity of the network alteration since crews deadheading on a flight, unlike the crews covering it, may not be required to connect with a follow-on.

The procedure to obtain an integer solution will repeatedly fix follow-ons, alter the long-haul network, generate new columns, and re-solve the LP until an integer solution is obtained. The procedure will always result in a feasible solution since deadheading is allowed.

8 COMPUTATIONAL EXPERIENCE

Our solution procedure was tested using two crew-assignment test problems, called *February* and *March*, from a long-haul operating airline. The problem characteristics (for the time-line network) are summarized in Table 2.1. The planning period considered is one month with the maximum TAFB equal to 15 days, resulting in a planning horizon of 45 days. Deadhead flights were selected using the procedure proposed by Barnhart *et al.* (1992). In this procedure, deadhead flights that can improve the solution are identified using a dual pricing procedure and are included in our network as deadheading opportunities.

The implementation, coded in Fortran uses the IBM Optimization Subroutine Library, Release 2 (OSL) (Druckerman *et al.* 1991) and runs on an IBM RS/6000 workstation.

Table 2.2 shows the quality of the crew-pairing solutions obtained with our column-generation-based procedure and compares them to the solutions obtained by the airline's schedulers. Pay-and-credit (column 2) is used by the airline to

Table 2.1 Test-problem characteristics.

Test problem	Number of stations	Number of crewbases	Number of flights	Number of deadheads	Number of duty periods	Number of nodes	Number of arcs
February	39	2	656	183	2981	5112	8061
March	41	2	833	397	5778	9645	15386

Table 2.2 Comparison of solutions for two test problems.

Test problems	Pay-and-credit LP/IP heuristic[†]	LP run[‡] time	IP run[‡] time	Airline solution
February	76.3/81%	2.22	0.32	108.1%
March	84.8/86%	7.9	0.22	107.1%

[†]First integer solution generated.
[‡]Time (hours) on an IBM RS/6000-540.

measure solution quality. It equals the difference of total pay (measured in equivalent number of hours of flying time) and operational flying time in hours (a constant for each monthly flight schedule) divided by this constant. (The pay-and-credit values shown in Table 2.2 are scaled proportionally to disguise the true values.)

For February, an integer solution was achieved representing a reduction in pay-and-credit from 108.1% (the solution achieved by the airline's crew schedulers) to 81%. Representatives of the airline estimate that about $250 000 are saved annually for each percent reduction in pay-and-credit shown in Table 2.2.

On an IBM RS/6000-540, the time to determine a solution to the February test problem totalled about 2.5 hours, with about 350 000 columns generated in the LP solution process.

For March, an integer solution with 86% associated pay-and-credit (Table 2.2) was determined as compared to the solution achieved by the airline's crew schedulers with 107.1% pay-and-credit. The total execution time on an IBM RS/6000-540 was about 8 hours with over 1.2 million columns generated.

9 CONCLUSIONS AND FUTURE RESEARCH

A solution to the long-haul crew-assignment problem is determined by first solving the linear programming relaxation of the crew-assignment problem using column generation. New columns are efficiently generated using a simplified shortest-path procedure exploiting the time-based structure of the specially constructed long-haul network. An alternative network formulation is also presented. Two approaches, one basic and one involving column generation, for obtaining an integer solution are discussed. Proof of the concept is provided by computational experience obtained in solving two crew-assignment problems for a long-haul operating airline. Representatives of the airline estimate that about $5 000 000 will be saved annually using the crew-pairing procedure described in this paper together with the deadhead-selection procedure of Barnhart *et al.* (1992).

ACKNOWLEDGEMENTS

The first author wishes to thank the National Science Foundation, Grant No. DDM-9058074 and American Airlines Decision Technologies for support of this research effort.

REFERENCES

Ahuja, R., Magnanti, T., and Orlin, J. (1993) *Network Flows: Theory, Algorithms, and Applications*, Prentice-Hall Inc., Englewood Cliffs, New Jersey.

Anbil, R., Gelman, E., Patty, B., and Tanga, R. (1991a) Recent advances in crew-pairing optimization at American Airlines, *Interfaces*, **21**, 62–74.

Anbil, R., Johnson, E. L., and Tanga, R. (1991b) *A Global Optimization Approach to*

Crew Scheduling, Computational Optimization Center Working Paper COC-91-05, Georgia Tech, Atlanta, Georgia.

Arabeyre, J. P., Fearnley, J., Steiger, F. C., and Teather, W. (1969) The airline crew scheduling problem: A survey, *Transportation Science*, **3**, 140–163.

Baker, E. K., Bodin, L. D., Finnegan, W. F., and Ponder, R. J. (1979) Efficient heuristic solutions to an airline crew scheduling problem, *AIIE Transactions*, **11** (2), 79–85.

Ball, M., and Roberts, A. (1985) A graph partitioning approach to airline crew scheduling, *Transportation Science*, **19** (2), 107–126.

Barnhart, C., Hatay, L., and Johnson, E. L. (1992) Deadhead selection for the long-haul crew pairing problem, *Operations Research*, to be published.

Barutt, J., and Hull, T. (1990) Airline crew scheduling supercomputers and algorithms, *SIAM News*, **23**, 6.

Crainic, G. T., and Rousseau, J.-M. (1987) The column generation and airline crew scheduling problem, *INFOR*, **25** (2), 136–151.

Dantzig, G. B., and Wolfe, P. (1960) Decomposition principle for linear programs. *Operations Research*, **8**, 101–111.

Desrochers, M., and Soumis, F. (1989) A column generation approach to the urban transit crew scheduling problem, *Transportation Science*, **23** (1), 1–13.

Desrosiers, J., Dumas, Y., Desrochers, M., Soumis, F., Sanso, B., and Trudeau, P. (1991) *A Breakthrough in Airline Crew Scheduling*, Report G-91-11, GERAD, Montreal, Canada.

Dijkstra, E. W. (1959) A note on two problems in connection with graphs, *Numerische Mathematik*, **1**, 269.

Druckerman, J., Silverman, D., and Viaropulos, K. (1991) *IBM Optimization Subroutine Library Guide and Reference, Release 2*, Document Number SC23-0519-2, IBM, Kingston, NY.

Etschmaier, M. M., and Mathaisel, D. F. X. (1985) Airline scheduling: an overview, *Transportation Science*, **19** (2), 127–138.

Gershkoff, I. (1989) Optimizing flight crew schedules, *Interfaces*, **19** (4), 29–43.

Lavoie, S., Minoux, M., and Odier, E. (1988) A new approach for crew pairing problems by column generation with an application to air transportation, *European Journal of Operational Research*, **35**, 45–58.

Marsten, R. E., and Shepardson, F. (1981) Exact solution of crew scheduling problems using the set partitioning model: Recent successful applications, *Networks*, **11**, 165–177.

Minoux, M. (1984) Column generation techniques in combinatorial optimization: a new application to crew pairing problems, *XXIV AGIFORS Symposium, France*, 15–29.

Odier, E., Lascaux, F., Hie, H. (1983) Medium haul trip pairing optimization, *XXIII AGIFORS Symposium, Memphis, TN*, 81–109.

Rannou, B. (1986) A new approach to crew pairing optimization, *XXVI AGIFORS Symposium, England*, pp. 153–167.

Rubin, J. (1973) A technique for the solution of massive set covering problems with applications to airline crew scheduling, *Transportation Science*, **7** (1), 31–43.

Ryan, D. M., and Falkner, J. C. (1988) On the integer properties of scheduling set partitioning models, *European Journal of Operational Research*, **35**, 442–456.

Shepardson, F., Marsten, R. E. (1980) A Lagrangean relaxation algorithm for the two duty period scheduling problem, *Management Science*, **26** (3), 274–281.

Thiriez, H. (1969) *Airline Crew Scheduling: A Group Theoretic Approach*, Working Paper, R69-1 (Doctoral Dissertation, Department of Civil Engineering, Massachusetts Institute of Technology).

Cynthia Barnhart

Department of Civil and Environmental Engineering
Massachusetts Institute of Technology
Cambridge
MA 02139
USA

Ellis L. Johnson
IBM Watson Research Center
Mathematical Sciences Department
Yorktown Heights
NY 10598
USA

Ranga Anbil
American Airlines Decision Technologies
Dallas/Fort Worth Airport
TX
USA

Levent Hatay
Kader Sokak 20/2
Gazi Osman Pasa
06700 Ankara
Turkey

3
The Contractor-Allocation Problem: An Integer Programming Problem that Wasn't

H. P. Williams
University of Southampton, UK

ABSTRACT
A practical problem of subcontracting work at minimum total cost is described and formulated as an integer programming model. It is then shown how this model can be converted to that of a minimum-cost network-flow model. This demonstrates that it was not really an integer program but a particularly simple type of linear program which could be optimized quickly and efficiently.

1 INTRODUCTION

The problem discussed here, and described in Section 2, was first presented to the author as a 'difficult' problem which it would almost certainly be impossible to optimize routinely in a reasonable period of time. Currently a heuristic method of solution was being employed. There was, however, interest in seeing if this heuristic always produced an optimal solution (as the client organization believed it would).

Therefore an integer programming (IP) model was built and solved optimally on a mainframe computer. (While this approach was necessary in 1981 it would now be possible to solve the model easily on a PC.) The results of solving the model suggested it has a special structure which guaranteed an optimal solution as a linear program (LP). In Section 3 the model is defined and the solution discussed. Further investigation suggested that this model could be transformed into that of a minimum-cost network-flow model. The recognition of such models and the application of a transformation method to the model described here are given in

Optimization in Industry 2. Edited by T. A. Ciriani and R. C. Leachman

Sections 4 and 5. In Section 6 further theoretical and practical discussion is given on the problem and its extensions.

This problem was first described in Cheshire *et al.* (1984).

2 THE CONTRACTOR-ALLOCATION PROBLEM

A large public-utility company subcontracts work on the basis of tenders. For the (fairly typical) instance of the problem being considered there were 28 subcontractors. Contractors employ a number of teams of men (typically a team consists of two men). Some contractors were large, employing over 100 teams, while others were much smaller and employed only a few teams.

Each subcontractor can bid for different types of work in each of 16 districts. From their information, and the known typical mix of work in each district, the company can obtain an estimated cost of employing teams from each contractor in such a district. Once allocated to a district a team has to carry out all necessary work.

Knowing the size of each contractor and the volume of work in each district the problem of allocating teams to districts at total minimum expected cost would be an example of a transportation problem. As such it would be an easy problem to solve.

There were, however, two further conditions which destroyed the transportation structure. First, in each district, it was stipulated that more than one contractor must be represented in order to avoid overdependence. The stipulation varied from district to district and the number of contractors was normally two or three. Secondly, at least one 'experienced' contractor had to be represented in each district in case of difficult work, or work involving safety considerations.

It was these two extra stipulations that persuaded the company to use a heuristic. This heuristic is described in Cheshire *et al.* (1984). In addition the planning process required a system that could be solved quickly on a minicomputer. Contractor allocations were made annually at a meeting involving district managers. Discussions took place, 'what if' questions were asked, and a number of computer runs took place before an acceptable solution was produced. At that time it would apparently have been impossible to solve and resolve an optimization model in the course of the planning meeting.

The total yearly costs of employing subcontractors was considerable (about £15 million in 1981 prices). Furthermore, the company was only one of 12 similar regional companies in the UK who had this, or a similar, problem. Hence there was great value in obtaining or demonstrating an optimal solution.

For the year considered, 28 contractors tendered with a total of 468 teams available. The total requirement over the 16 districts was for 334 teams. Ten of the contractors were regarded as experienced.

3 AN INTEGER PROGRAMMING MODEL

Data C_{ij} = expected yearly cost of allocating a team from contractor i to district j.

A_i = number of teams available from contractor i.

R_j = requirement of district j for teams.
N_j = minimum number of contractors to be represented in district j (2 or 3).

Variables x_{ij} = number of teams allocated from contractor i to district j.

$$\delta_{ij} = \begin{cases} 1 \text{ if contractor } i \text{ is represented,} \\ 0 \text{ otherwise.} \end{cases}$$

Objective Minimize $\sum_{i,j} c_{ij}x_{ij}$. (1)

Constraints

$$\sum_j x_{ij} \leqslant A_i \quad \text{for all } i, \tag{2}$$

$$\sum_i x_{ij} = R_j \quad \text{for all } j, \tag{3}$$

$$x_{ij} - \delta_{ij} \geqslant 0 \quad \text{for all } i,j, \tag{4}$$

$$\sum_i \delta_{ij} \geqslant N_j \quad \text{for all } j, \tag{5}$$

$$\sum_{\substack{i \in \text{experienced} \\ \text{contractors}}} \delta_{ij} \geqslant 1 \quad \text{for all } j. \tag{6}$$

The δ_{ij} variables are restricted to the integer values 0 and 1.

The x_{ij} variables would normally be regarded as continuous and rounded to their nearest integer values. They could of course be stipulated to be general integer variables if so required.

Since not every contractor tendered for work in each district the resultant model had 864 variables and 508 contraints. It was considered as a mixed or pure integer programming model depending on whether the x_{ij} variables were regarded as integer.

In order to illustrate the model clearly we give the formulation in the modeling language MAGIC (see Day and Williams, 1986):

```
NAME:ALLOCAT;                                              {MODEL NAME}
NCONT=28;                                         {NUMBER OF CONTRACTORS}
NDIST=16;                                            {NUMBER OF DISTRICTS}
NUMEXP=10;                          {NUMBER OF EXPERIENCED CONTRACTORS}
COST(NCONT:NDIST)=COSTFILE;                            {COST DATA FILE}
TEAMS(NCONT)=TEAMFILE;                        {TEAM AVAILABILITY FILE}
DEMAND(NDIST)=DEMFILE;                              {TEAM DEMAND FILE}
NUMB(NCONT:NDIST);                  {LP VARIABLE FOR NUMBER OF TEAMS
                              FOR EACH CONTRACTOR FOR EACH DISTRICT}
IND(NCONT:NDIST) INTEGER;   {INDICATOR VARIABLE TO SHOW IF ANY TEAMS ALLOCATED}
TOTCOST(MIN):SIGMA I=1,NCONT:SIGMA J=1, NDIST:
                    COST(I,J)*NUMB(I,J):              {OBJECTIVE FUNCTION}
AVAIL(I=1,NCONT):SIGMA J=1,NDIST:NUMB(I,J)<=TEAMS(I); {AVAILABILITY
                                                         CONSTRAINTS}
FOR J=1,NDIST;
  REQUI(J):SIGMA I=1,NCONT: NUMB(I,J)=DEMAND(J); {REQUIREMENT CONSTRAINTS}
    FOR I=1,NCONT;
      LINK(I,J):NUMB(I,J)-IND(I,J)>=0;              {LINKING CONSTRAINTS}
    END;
  MIN(J):SIGMA I=1,NCONT: IND(I,J)>=3;  {MINIMUM NUMBER OF CONTRACTORS
                                          PER DISTRICT}
  EXP(J):SIGMA I=1,NUMEXP:IND(I,J)>=1;  {AT LEAST ONE EXPERIENCED CONTRACTOR
                                          PER DISTRICT}
END;
MPS ALLOFILE;     {PRODUCE A FILE OF INPUT DATA IN MPS FORMAT}
STOP;
```

This model was solved using the MPSX (Mixed Integer Programming) package on an IBM 370/165. (Today it would be possible to solve it on a PC.) While it was intended to use the branch-and-bound algorithm it turned out that an integer solution was obtained to the LP relaxation. This property was maintained when the right-hand-side and objective coefficients were changed. The result was that what seemed at first sight to be a difficult IP model appeared to be an LP model and therefore optimizable in a modest time (34 seconds were needed—today it would be solvable in a few seconds on a PC).

It is well known that some models (such as the transportation problem) produce optimal LP solutions which are integer. Therefore it seemed worth investigating further as to whether the model had such a property.

4 TOTALLY UNIMODULAR AND NETWORK-FLOW MODELS

A matrix A is said to be totally unimodular if each square submatrix has determinant 0 or ± 1. By a result due to Veinott and Dantzig (1968) it is straightforward to show that this guarantees that an LP model with constraints

$$Ax = b$$

has integral extreme points for all integral b. Unfortunately it is not computationally practical to test an arbitrary matrix for total unimodularity in this way. There are, however, easily detected properties which guarantee total unimodularity (see, for example, Nemhauser and Wolsey, 1988, or Williams, 1990). The matrix of the model described above does not obviously possess any of these properties.

Many practical instances of totally unimodular models arise as minimum-cost network-flow (MCNF) models. Although the class of models with totally unimodular matrices is larger than that of MCNF models it seemed worthwhile to investigate whether this model could be transformed into an MCNF model.

A number of instances of the successful conversion of a model to the MCNF form are reported in the literature: for example, Veinott and Wagner (1962), Daniel (1973), Wilson and Willis (1983), and Wagner (1969). The general applicability of the MCNF model has been discussed by Glover and Klingman (1977) and Bradley (1975). Seymour (1980) characterizes the relationship between MCNF models and the wider class of totally unimodular models.

5 CONVERTING LP MODELS TO MCNF MODELS

One of the best-known methods of converting (if possible) an LP model to an MCNF model is due to Bixby and Cunningham (1980). Another method is due to Iri (1966).

We chose to use the method described by Baston *et al.* (1991). Full details are given in that paper. The method considers the matrix A in a 'standard' form in which it contains an identity portion. For illustration we will consider the matrix given below:

	1	2	3	4	5	6	7	8	9	10
1	1					1				1
2		1				1	1			1
3			1				1	1	1	1
4				1			1		1	1
5					1	−1		1	1	

The columns will correspond to arcs of the network being sought. The columns of the identity portion have a one–one correspondence with the rows which therefore also correspond to arcs. These arcs will provide a spanning tree for the network. In order for the matrix to represent a (undirected) network the other columns must correspond to arcs which span those arcs in the spanning tree corresponding to arcs with nonzero entries in the column. For example, in Figure 3.1 we have the arc corresponding to column 6 spanning arcs 1, 2, and 5 of the spanning tree. Other columns in the nonidentity portion are similarly represented in Figure 3.1.

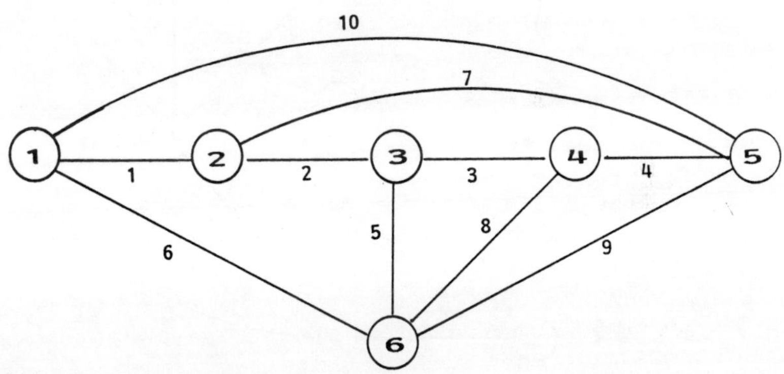

Figure 3.1

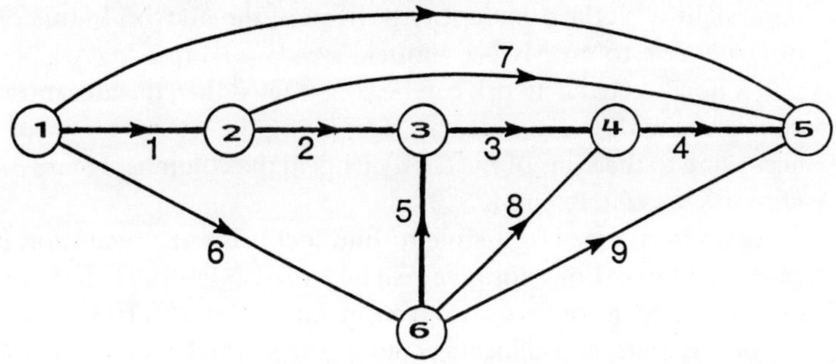

Figure 3.2

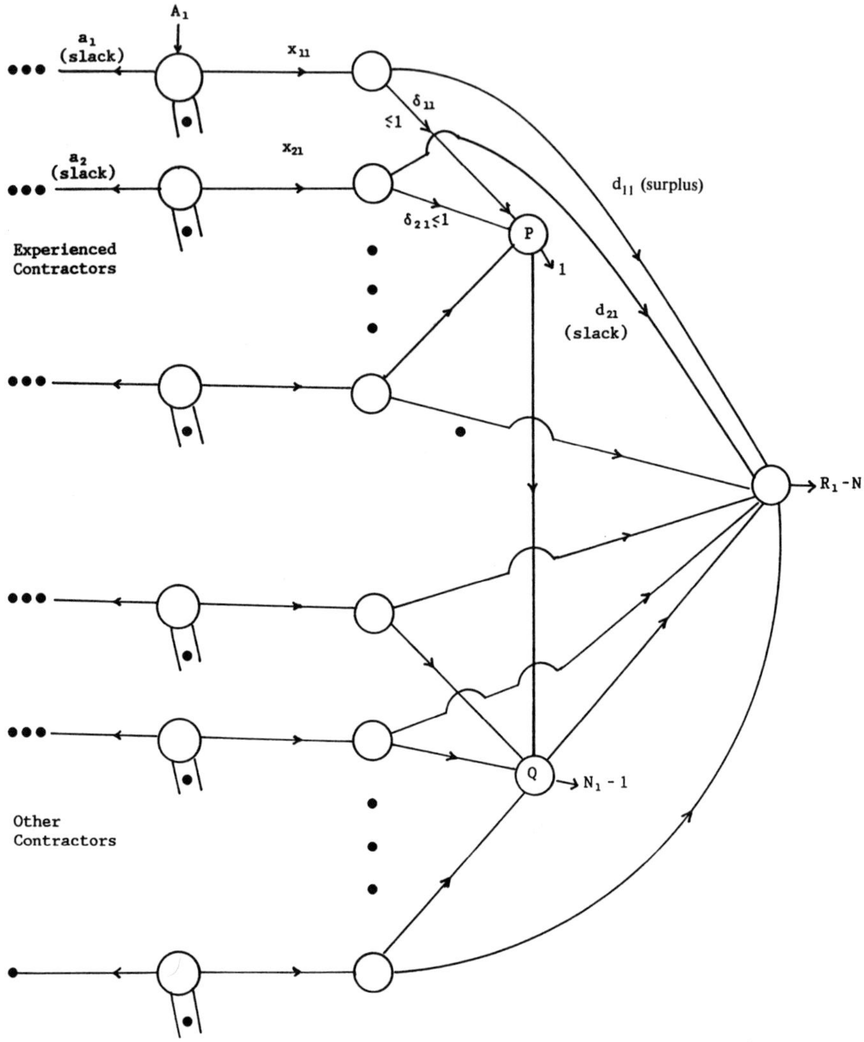

Figure 3.3

It may well be impossible to configure the arcs in the spanning tree to be consistent with the nonidentity portion of the matrix. In this case the matrix does not give rise to an MCNF model.

If a nondirected network can be constructed then the arcs must be given directions consistent with the nonidentity arcs being the 'vector sums' of spanning-tree arcs according to the signs of the coefficients in the columns. Figure 3.2 gives the directed network for the example.

Again it may be impossible to find a consistent orientation for the arcs; in this case the matrix does not give rise to an MCNF model. Examples to illustrate the different circumstances are given in Baston *et al.* (1991).

For the contractor-allocation model the method produces the MCNF model given in Figure 3.3. Only a portion of the network is drawn in this figure relating to district 1. The a_i are the slack variables in constraints (1) and d_{ij} are the surplus variables in constraints (4). The arcs δ_{ij} have upper bounds of 1. Since node P has an outward flow of 1 and arc Q an outward flow of $N_1 - 1$, there is effectively a backward flow of 1 along PQ. To overcome this flow at least one unit (team) must flow along PQ. Since arcs δ_{ij} all have upper bounds of 1 this forces at least one experienced contractor to be used. Similarly the outward flow of $N_1 - 1$ from Q forces at least $N_1 - 1$ other contractors to be represented in addition to the experienced contractor.

In Cheshire *et al.* (1984) a slightly different network was obtained by other means using a simple lower bound of 1 on the equivalent of arc PQ.

The resultant model had 77 nodes and 912 arcs (not all contractors tendered for all districts). Such a model is of very modest size and can be solved very rapidly on specialist software (see e.g. Bradley *et al.*, 1977). Such software uses specialist data structures and procedures to represent and manipulate networks and trees with immense savings in time and space.

Another advantage of this approach is that networks are usually more easily understood by managers than general LP or IP models.

The resultant model could be repeatedly solved optimally on a small computer leading to a saving of about 2% over the heuristic solution.

Having demonstrated that the model was not really an IP but an LP, meaningful dual values could be attached to the constraints (5). The stipulation as to whether a particular district needed two or three contractors represented was rather arbitrary. The dual values allowed this stipulation to be costed.

There was some worry that by optimizing some small contractors who would otherwise survive might be driven out of business. For the particular instance of the problem considered it turned out that exactly the same contractors were so driven out as had been from the heuristic.

It should be remarked that some other regional companies had a variant of the problem in which there was an *upper* limit on the number of contractors represented in each district. This variant results in a genuine (although fairly easy) IP model.

REFERENCES

Baston, V. J. D., Rahmouni, M. K., and Williams, H. P. (1991) The practical conversion of linear programmes to network flow models, *European Journal of Operational Research*, 50, 325–334.

Bixby, R. E., and Cunningham, W. H. (1980) Converting linear programs to network problems, *Mathematics of Operations Research*, 5, 321–357.

Bradley, G. H. (1975) Survey of deterministic networks, *AIIE Transactions*, 7, 222–234.

Bradley, G. H., Brown, G. G., and Graves, G. W. (1977). Design and implementation of large scale primal transhipment algorithms, *Management Science*, **24**, 1–34.

Cheshire, M., McKinnon, K. I. M., and Williams, H. P. (1984) The efficient allocation of private contractors to public works, *Journal of the Operational Research Society*, 35, 705–709.

Daniel, R. C. (1973) Phasing out capital equipment, *Operational Research Quarterly*, **24**, 113–116.

Day, R. E., and Williams, H. P. (1986) MAGIC: The design and use of an interactive modelling language for mathematical programming, *IMA Journal of Mathematics in Management*, **1**, 53–65.

Glover, F., and Klingman, D. (1977). Network applications in industry and government, *AIIE Transactions*, **9**, 363–376.

Iri, M. (1966) *A Criterion for the Reducibility of a Linear Programming Problem to a Network Flow Problem*, RAAG Research Notes, Third Series, No. 98.

Nemhauser, G. L., and Wolsey, L. A. (1988) *Integer and Combinatorial Optimization*, John Wiley & Sons, New York.

Seymour, P. D. (1980) Decomposition of regular matroids, *Journal of Combinatorial Theory, B*, **28**, 305–359.

Veinott, A. F., and Dantzig, G. B. (1968) Integral extreme points, *SIAM Review*, **10**, 371–372.

Veinott, A. F., and Wagner, H. M. (1962) Optimal capacity scheduling—I, *Operations Research*, **14**, 45–51.

Wagner, H. M. (1969) *Principles of Operations Research*, Prentice-Hall, Englewood Cliffs, NJ.

Williams, H. P. (1990) *Model Building in Mathematical Programming*, John Wiley & Sons, Chichester.

Wilson, E. J. G., and Willis, R. J. (1983). Scheduling of telephones betting operators—A case study, *Journal of the Operational Research Society*, **34**, 991–998.

Faculty of Mathematical Studies
University of Southampton
Southampton
SO9 5NH
UK

4
The Space-Module Payload-Accommodation Problem: An MIP Formulation

Giorgio Fasano
Alenia Spazio S.p.A., Turin, Italy*

ABSTRACT
A challenging task of space-module design is that of payload accommodation. The problem consists of the allocation of payloads to racks, within the modules, taking into account both the geometrical constraints and the on-board resource availability. A specific software tool aimed at performing the space-module accommodation task is presented in this paper. It is based on the mixed-integer linear programming (MIP) techniques and is oriented to provide quick solutions that guarantee efficient utilization of the resources and at the same time meet payload requirements.

1 INTRODUCTION

An important task of the space-module design is the allocation of items (payloads) to module racks. Recently Alenia Spazio has analyzed the payload-accommodation aspects of the Attached Pressurized Module of the Space Station Freedom (see Fasano *et al.*, 1992).

The possible payload allocations in racks within the module are determined by geometrical constraints and by different resource availability for the rack locations. A payload consists of a set of facilities which support experiments with defined

*Alenia Spazio SpA is a Company of Alenia SpA, the new Company belonging to the IRI-Finmeccanica Group, resulting from the merger of Aeritalia and Selenia, taking over full responsibility for their functions, responsibilities, and activities.

Optimization in Industry 2. Edited by T. A. Ciriani and R. C. Leachman

resource requirements. The operations of these experiments must satisfy the system constraints. The problem of payload allocation in racks is complex and requires a specific software tool to efficiently perform the payload-accommodation task. This tool should be able to identify bottlenecks, evaluate different operational options, and optimize the use of the available module resources.

2 GENERAL ASPECTS

Even if the optimal solutions were not an explicit requirement, the mathematical programming approach seemed to be quite suitable for the purpose.

In particular the nature of the problem and the research results obtained in recent years (see, for instance, Anbil *et al.*, 1992), as well as the recently improved optimization techniques (see Johnson *et al.*, 1992), encouraged us to use a mixed-integer programming (MIP) formulation.

Note

A linear mixed-integer programming problem P can be formulated (see Johnson and Nemhauser, 1992) as min $\{cx + hy\colon Ax + Gy \leqslant b, x \in Z^n_+, y \in R^p_+\}$ where Z^n_+ is the set of non-negative integral n-dimensional vectors, R^p_+ is the set of non-negative real p-dimensional vectors, $x = (x_1, \ldots, x_n)$ and $y = (y_1, \ldots, y_p)$ are the variables.

The set $S = (x \in Z^n_+,\ y \in R^p_+,\ Ax + Gy \leqslant b)$ is called the *feasible region*, and $(x,y) \in S$ is called a *feasible solution*. The function $z = cx + hy$ is called the *target function*. A feasible point (x^O, y^O) for which the target function is as small as possible, that is $cx^O + hy^O \leqslant cx + hy$ for all $(x,y) \in S$ is called an *optimal solution*.

The payload-accommodation software tool is based on different MIP mathematical models and takes into account the two following goals:

- identification of operational limits (bottlenecks), in order to suggest different and more suitable on-board resource distributions;
- search for satisfactory 'dynamic' (time-dependent) solutions, i.e. feasible payload accommodations, verified against the operational constraints. This task, at present under development, will provide preliminary time-lines. The payload accommodation obtained in such a way serves as a starting point for further optimizations, entirely dedicated to the scheduling aspects.

A pairing-based approach, in a certain sense analogous to that achieved for the 'crew-pairing optimization' problem (see Anbil *et al.*, 1992) accommodates both goals.

The rack-level operational-status generator program generates the feasible operational scenarios. One operational scenario consists of a group of experiments of the same payload that can be operated together within a rack (rack-level operational status). Another program, the system-level operational-status generator program, generates a set of well-sampled operational scenarios, in which pairings of different racks can work together (system-level operational status).

The model constraints deal with both geometrical and operational (resource-related) conditions. The former are essentially related to the assignment aspects of the problem (allocation of each payload to a rack) under specific requirements. Present models take into account the following resource availability at both local and system levels:

- electrical power,
- heat rejection, i.e. water cooling and air cooling,
- crew time.

The model's unknowns are essentially those related to the payload allocation (0–1 variables) and to the number of cycles per experiment (generally approximated by continuous variables). The time duration, associated with both rack-level and system-level operational status is another variable. Each payload facility is associated at least to an operational status within which the corresponding experiment can be operated (together with the other possible experiments associated with the same operational status).

Another experiment variable, for each associated operational status, takes into account the number of cycles to be operated within the operational-status period. Some operational-status durations can be zero: in this case the facilities involved will never work in the corresponding operational configuration. Similarly, if some variables, relative to the number of cycles within an operational status are zero then only a subset of the operational-status interrelated experiments will operate together, for some instant, within the operational-status period.

Concerning the rack-level operational status, each payload will be working at any instant in a specific operational configuration, consisting of a subset of contemporarily active facilities (included the empty subset, in which no experiment is active). Each operational configuration is related to (at least) one operational status during the payload 'operational life', i.e. the payload-activity duration is partitioned by the set of operational statuses. As a consequence the sum of the operational-status durations must not exceed the global time availability, i.e. the time increment. Analogous considerations apply also with the system-level operational status. In this case (for reasons due to the numerous possible scenarios), only a selection from all possible scenarios is considered. The system-level operational statuses represent a partition of the resulting (reduced) system 'operational life'.

As in the rack-level case, the sum of the operational-status durations must not exceed the time increment. The system-level and rack-level operational status duration, as well as the number of cycles per experiment for each operational status, represent the time-dependent aspects of the payload accommodation 'dynamic' approach. To understand the operational-status concept better, two examples will be given.

Example 1—operational status at rack level

Consider a payload containing three facilities, each one dedicated to a specific experiment. Suppose that the payload activity requires just the electrical power

Table 4.1 Payload energy-consumption data.

	Electrical-power consumption (per instant)
Facility 1	3 kW
Facility 2	3 kW
Facility 3	2 kW

as the only resource involved and also assume that the power consumption is constant during each experiment cycle for any facility. Finally assume that the power available (at any instant) for the rack R is 6 kW. The data is shown in Table 4.1.

The rack-level operational status can be derived. If the payload is allocated in the rack R, then all the feasible operational statuses are represented as shown in Table 4.2. The first line means that within the operational status 1 only facility 1 can be activated; the fifth line means that within the operational status 5, facilities 1 and 2 can work together for some instant; and so on.

Example 2—operational status at system level

The system-level operational-status concept arises when a suboptimal and preliminary timeline has to be defined. Suppose that the allocation of each payload is known. Then the goal becomes that of determining valid and efficient resource exploitation (with respect to a specific optimization criterion), compatible with the given allocations. Since the allocation for each payload is known, the upper-limit resource consumption for each rack can be easily determined.

Referring to Example 1, if for instance the payload considered is allocated in the rack R, the maximum electrical-power consumption for the rack under consideration is 6 kW. A 'worst-case' scenario may also be derived, corresponding to a maximum resource consumption at any instant of activity, for each payload. The system operational statuses are thus introduced to generate a set of well-sampled system operational configurations, within which different payloads can work together, i.e. without exceeding the system per-instant resources availability.

It should be noticed that, for instance, the operational status 5 (Table 4.2) includes the operational statuses 1 and 2. As a consequence, it is clear that a subset of the feasible operational statuses becomes redundant. The scope of the rack-level operational-status generator program is to determine all the nonredundant

Table 4.2 Payload feasible operational status in R.

	Facility 1	Facility 2	Facility 3	Operational-status maximum power consumption
Operational status 1	X			3 kW $\leqslant$ 6 kW
Operational status 2		X		3 kW $\leqslant$ 6 kW
Operational status 3			X	2 kW $\leqslant$ 6 kW
Operational status 4	X		X	5 kW $\leqslant$ 6 kW
Operational status 5	X	X		6 kW $\leqslant$ 6 kW
Operational status 6		X	X	5 kW $\leqslant$ 6 kW

operational statuses for all the payloads. On the other hand, the scope of the system-level operational-status generator program is to determine a set of well-sampled nonredundant system-level operational statuses.

3 MATHEMATICAL MODEL

A mathematical model formulation has been achieved to treat all the different problems (bottleneck identification, search for satisfactory accommodations as well as preliminary and suboptimal time-lines). It is based on a parametric structure, such that each particular model corresponds to a specific parameter setting. Each model is automatically generated by means of the model-generator program.

In the present section, the general mathematical model formulation is illustrated, pointing out, when necessary, some peculiarities of specific models.

Strictly speaking, the variables involved are:

- nonnegative continuous variables (operational-status duration and resources consumption),
- 0–1 variables (payload allocation and other decisional aspects),
- nonnegative integer variables (number of cycles).

However, when the range between the integer variables' upper and lower limit is quite wide, they are obviously considered as continuous. With this preliminary statement, the general mathematical-model structure is illustrated below (in a simplified form).

For each payload (i) and for each rack (j), a set of 0–1 variables x_{ij} is introduced with the following meaning:

$x_{ij}=1$ if, and only if, the payload (i) is allocated to the rack (j)
$x_{ij}=0$ otherwise.

The following two kinds of equations represent the assignment aspect of the problem:

- each payload must be allocated to one rack and vice versa,
- each rack must contain one (and only one) payload.

$$\forall i \sum_j x_{ij}=1. \tag{1}$$

$$\forall j \sum_i x_{ij}=1. \tag{2}$$

A set of constraints, involving some of the x variables, is introduced. The constraints represent geometrical conditions for some payload subsets but for simplicity are not treated in the present chapter.

For each payload (i), each contained facility (h) and each rack (j), the (continuous) variable n_{ihj} is considered. It represents the number of cycles to be done by the facility (h) of the payload (i) in the rack (j). Of course n_{ihj} must be zero for each (j), except for that one corresponding to the rack to be chosen for the payload assignment. This fact is expressed by the following inequalities:

$$\forall(i,h,j) \qquad n_{ihj} \leqslant T_{ih} x_{ij}, \tag{3}$$

where the terms T_{ih}, are the maximum number of cycles required for the facility (h). Moreover, a minimum number of cycles is required for each facility (h). This is obtained by means of the following inequalities:

$$\forall(i,h) \qquad \sum_j n_{ihj} \geqslant N_{ih}, \tag{4}$$

where the terms N_{ih} represent the minimum number of cycles.

If the rack-level operational status s is applied to the facility (h) of (i), the (continuous) variable r_{ihs} is inserted. This represents the number of cycles to be done by the facility (h) within the operational status s of the payload (i). The conditions (5) noted below apply, meaning that the total number of cycles to be done by the facility (h) must be equal to the sum of all cycles to be done within each operational status.

$$\forall(i,h) \qquad \sum_j n_{ihj} = \sum_s r_{ihs}. \tag{5}$$

The variable z_r $(r=1,2, \ldots)$, representing the related duration is associated with each system-level operational status r generated.

The sum of all durations cannot exceed the global time increment Z. This fact is expressed by the following inequality:

$$\sum_r z_r \leqslant Z. \tag{6}$$

The continuous variables z_{irs} are also introduced: they represent the duration of the payload (i) rack-level operational status s, within the system-level operational status r. The quantity

$$z_{is} = \sum_{r \in R^*_{is}} z_{irs}$$

represents, consequently, for each payload (i), the duration of the (rack-level) operation status s. The subset $R^*_{is} \subseteq R$ includes all indices r whose corresponding (system-level) operational statuses allow the (rack-level) operational status s for the payload (i).

Since the rack-level operational-status feasibility depends on the allocation assumed by the payload (certain operational statuses are feasible only in specific rack types, see rack-level operational-status definition and Example 1), the following relationship between the variables z_{is} and the decisional ones x_{ij} is stated:

$$\forall(i,s) \qquad \sum_{r \in R^*_{is}} z_{irs} \leqslant Z \sum_{j \in J^*_{is}} x_{ij}. \tag{7}$$

The subset $j^*_{is} \subseteq J$ includes all indices j whose corresponding racks allow the operational status s for the payload (i). When (i) is allocated in a rack (j), for which the operational status s is feasible, the inequality (7) becomes redundant. Indeed, in this case, $x_{ij} = 1$ and, as a consequence,

$$z_{is} = \sum_{r \in R^*_{is}} z_{irs} \leqslant Z.$$

Otherwise, all the variables x_{ij}, corresponding to the allocations for which the status s is feasible, are zero.

In this case the inequality (7) implies $z_{is} = 0$ and the operational status s does not exist, since its duration z_{is} is zero.

Analogous considerations should be made when dealing with the system-level operational status. In this case each term $\sum_{s \in S^*_{ir}} z_{irs}$ corresponds to the global duration of the payload (i) activity within the system-level operational status r. The subset $S^*_{ir} \subseteq S$ includes all indices s whose corresponding (rack-level) operational status s for the payload (i) is admissible within the (system-level) operational status r.

The set $J^{**}_r \subseteq J$ contains all the indices j of the racks that can work together in the state r. The following inequalities express these conditions:

$$\forall(i,r) \qquad \sum_{s \in S^*_{ir}} z_{irs} \leqslant Z \sum_{j \in J^{**}_r} x_{ij}. \tag{8}$$

Moreover, if the rack-level operational-status duration z_{is} is zero, the number of cycles to be done by the corresponding facility (h) must also be zero.

The following conditions hold:

$$\forall(i,h,s) \qquad r_{ihs}\, Z_{ih} \leqslant \sum_{r \in R^*_{is}} z_{irs}, \tag{9}$$

where the term Z_{ih} represents the cycle duration of the facilities (h).

The duration of the payload (i) activity within the system-level operational status r cannot exceed the period z_r. These conditions are given by the inequalities:

$$\forall(i,r) \qquad \sum_{s \in S^*_{ir}} z_{irs} \leqslant z_r. \tag{10}$$

The resources involved (up to now) in the mathematical formulation are limited to the electrical power, the heat rejection (water plus air) and the crew time. Notice that the aspects dealing with the instantaneous resource availability, with the simplifying limitations already illustrated when treating the models 'dynamic' characteristics are implicitly included in the operational-status concepts. Apart from the resource related to the crew time, the two aspects below are to be noted:

- local (per-rack) resource availability throughout the time increment,
- global resource availability during all the time increment.

In the present chapter the problems of water cooling (which is quite particular) and electrical power are illustrated. The other resources, being quite similar to the last one, are not treated. The local availability of electrical power is represented by the following conditions:

$$\forall j \qquad \sum_{i,h} n_{ihj}\, E_{ih} \leqslant E_j. \tag{11}$$

For each (i,h) the E_{ih} terms are the electrical-energy consumption per cycle. The E_j terms are the maximum energy that can be exploited by the rack (j). Of course, depending on the different options, the value of E_j can be the peak level or the nominal; or it can be limited to a specific percentage. Notice that in general the sum of all local-energy availability is greater than the global system quantity; i.e. generally only a subset of racks can work at the maximum level. The conditions (12) are introduced to express the global electrical-energy availability:

$$\sum_{i,h,j} n_{ihj}\, E_{ih} \leqslant E. \tag{12}$$

The meaning of the coefficients is the same as before and similar remarks hold.

The target function considered (up to now) has been to maximize the sum of all percentages of cycles (with respect to the maximum number required per experiment). It is expressed by:

$$\max \sum_{i,h,j} \frac{n_{ihj}}{T_{ih}}. \tag{13}$$

Other target functions are also under study, dealing for instance with the mimization of the water-resource waste.

4 ALGORITHMIC PERFORMANCE

A set of different models has been analyzed. Some results are illustrated, giving particular attention to the dimensions and general characteristics of the models, as well as to the CPU time consumption. The IBM Optimization Subroutine Library optimizer has been used on an IBM RISC/6000 560 platform.

Even though the model structure (see equations (1) and (2)) seems quite suitable to be treated by using the special ordered sets of variables (SOS, see Williams,

1990), it appeared advantageous to follow a different approach, based on a specific branching order. This consists of processing the x variables in a sequence corresponding to the assignment of one payload at a time, starting from those for which the more restrictive requirements are given. The optimizer preprocessor (see IBM, 1992) serves to reduce the CPU time consumption consistently.

A set of cases dealing with the Space Station Attached Pressurized Module (see Fasano *et al.*, 1992) is reported here. The number of payloads involved is 21. They have been analyzed to identify the operational bottlenecks. Optimality has always been reached.

Case 1

- Resource availability:
  one crew member, all other resources 100%.
- Model dimensions:
  rows: about 8000,
  columns: about 9000,
  0–1 variables: about 350.
- CPU time consumption:
  minutes to reach optimality: about 10.

Case 2

- Resource availability:
  one crew member, 8 hours a day, can operate up to 5 experiments at a time; all other resources 100%.
- Model dimensions:
  rows: about 8000,
  columns: about 9000,
  0–1 variables: about 350.
- CPU time consumption:
  minutes to reach optimality: 10.

Case 3

- Resource availability:
  one crew member, 8 hours a day, can operate up to 3 experiments at a time; all other resources 50%.
- Model dimensions:
  rows: about 8000,
  columns: about 9000,
  0–1 variables: about 350.
- CPU time consumption:
  minutes to reach optimality: 60.

5 CONCLUSIONS

The main result of the software tool is to obtain a set of feasible payload accommodations, each of them able to satisfy the payload requirements, with different performances, depending on the considered geometrical and operational constraints. The possibility of analyzing several options guarantees flexibility and quick response any time that different operational constraints need to be considered and allows the evaluation of the impacts on the payload accommodation any time that a system design change varies the resource available to payloads. Further developments oriented toward both the mathematical models and the software programs, as outlined in this paper, are foreseen in order to cover other aspects, and thus achieve more complete payload-accommodation analysis.

REFERENCES

Anbil, R., Tanga, R., and Johnson, E. (1992) A global approach to crew pairing optimization, *IBM System Journal*, **31** (1).

Fasano, G., Piras, A., and Bosio, T. (1992) Columbus dynamic payload accommodation: Software models for pressurized modules in space, presented at the *IBM Europe Institute Oberlech (Austria), 1992.*

IBM (1992) *Optimization Subroutine Library Guide and Reference, SC23-0519*, IBM Corporation.

Johnson, E., and Nemhauser, G. (1992) Recent development and future directions in mathematical programming, *IBM System Journal*, **31** (1).

Williams, H. P. (1990) *Model Building in Mathematical Programming*, John Wiley & Sons, Chichester.

Alenia Spazio S.p.A.
Corso Marche 41
10145 Turin
Italy

5
The Forward–Reserve Problem

E. H. Frazelle, S. T. Hackman, U. Passy and L. K. Platzman
Georgia Institute of Technology, GA, USA

ABSTRACT
Many distribution centers are configured with a forward area for broken and full case picking and a reserve area which stores incoming receipt of stock. Since the picking cost in the reserve area is much higher than the picking cost in the forward area, overall picking costs can be reduced by assigning items to the forward area. However, picking productivity decreases as the size of the forward area grows.

A critical design decision is the size of the forward area. To determine the appropriate size it is necessary to solve the assignment–allocation subproblem, namely, to determine which items should be assigned to the forward area, and in what quantities. In this chapter we formulate and develop two algorithms to solve the forward–reserve problem. We prove that items may be ordered *a priori* according to a simple ratio which depends only on the characteristics of the items and is independent of the warehouse parameters. The ratio, which we call the *Economic Assignment Quotient*, transforms the assignment–allocation subproblem into a simple ranking problem. A statistical approach is discussed and a case study is provided.

1 INTRODUCTION

A recent survey of warehousing professionals identified order picking as the warehousing activity in most need of engineering and management attention (WERC, 1986). The sense of urgency arises out of several motivations. First, order picking is the most costly activity in a typical warehouse, representing nearly 65% of the total costs and 50% of the workforce (WERC, 1986). Second, new marketing and distribution policies including micro-marketing, cycle-time reduction, and quick response have increased the demands on the order-picking activity. Specifically, faster, more accurate transactions are required with less inventory of more

Optimization in Industry 2. Edited by T. A. Ciriani and R. C. Leachman

Figure 5.1 Typical forward area for broken case picking.

stock-keeping units. The conventional response to these increased requirements is to employ more labor or to invest in more automated equipment. This may not be always feasible due to labor shortages in many areas and to high hurdle rates brought on by short-term business initiatives. In this chapter we develop a stock assignment–allocation policy that will reduce the overall work content in order picking. In industrial parlance the problem we shall consider is known as the *forward–reserve problem* (FRP).

Many distribution centers are configured with a *forward area* for broken-case and full-case picking and a *reserve area* for pallet picking and reserve storage. A typical forward-area layout includes bin shelving and gravity-flow rack arranged in an aisle configuration (see Figure 5.1). A typical reserve-area layout includes pallet rack arranged in an aisle configuration and accessed with lift trucks or automated storage/retrieval machines. Upon initial receipt all items are assigned a location in the reserve area. A typical receipt quantity ranges from several cases to several pallets. If an item is assigned to the forward area, all picks for that item are executed in the forward area. The picks for items not assigned to the forward area take place in the reserve area. When the inventory of an item assigned to the forward area reaches its restocking level, an *internal replenishment* is executed. An amount of stock corresponding to the size of the location in the forward area is retrieved from the reserve area, transported, and stocked in the forward area.

A critical design decision is the amount of space to allocate in the forward area for each item. The picking cost in the reserve area is much higher than the picking cost in the forward area, and so overall picking costs can be reduced by assigning items to the forward area. When more items are assigned to the forward area, it grows in size. As the size of the forward area increases, the picking productivity decreases, leading to an increase in the cost to pick in the forward area. Eventually the size of the forward area will become too large, thereby defeating the whole point of having a forward area in the first place. One can reduce the size of the forward area, with no increase in the number of items, but this will necessarily increase the number of internal replenishments. Typically, the cost of an internal replenishment is several times the cost of a pick.

The economic trade-offs in designing the optimum size of the forward area discussed above have been documented in the literature (see, for example, White and Kinney, 1982). Bozer (1985) developed a quantitative model that explictly accounts for some of the economic trade-offs. His model concentrates on the actual location of items within the forward area, since he assumes that the items assigned to this area have been predetermined. In this chapter we develop an aggregate model that determines which items should be assigned to the forward area, and in what quantity. We also explicitly model how the size of the forward area affects the picking costs.

In Section 2 we provide a formal definition of the (forward–reserve) problem for a cart picking system. In Section 3 we analyze the mathematical structure of the FRP model. We prove that the items may be ranked *a priori* according to a simple ratio—independent of problem parameters—as to their desirability for assignment in the forward area. We call this ratio the *Economic Assignment Quotient* (EAQ). We use the EAQ to design a simple algorithm that solves the FRP to near-optimality. The EAQ was first discovered in Hackman and Rosenblatt's (1990) solution to an assignment–allocation subproblem that we will discuss more generally in Section 3. Section 4 develops and validates a modification to our algorithm based on sampling that will reduce the computational effort for large databases. Section 5 describes a simple case study from a mail-order distribution warehouse. Additional applications are briefly outlined in Section 6.

2 PROBLEM DEFINITION

The forward–reserve problem arises in a variety of warehouses. For example, the forward picking area could be the floor-level in conventional pallet rack for case-picking palletized unit loads, horizontal or vertical carrousels for small items, or even automated dispensing machines for piece-picking cosmetics, films, cassettes, or compact disks. There are also a variety of alternative configurations for reserve picking systems including block stacking, conventional pallet rack accessed by counterbalanced lift trucks, narrow-aisle systems accessed with reach or turret trucks, or automated storage/retrieval systems. Both in the formulation and in the case study we focus on a traditional forward–reserve configuration, which will make concrete the fundamental trade-offs and parameters used to derive forward–reserve solutions.

One of the most popular methods for order picking in a conventional warehouse is cart picking from static and/or gravity-flow rack (see Figure 5.2B). In a cart picking system, an order picker is assigned a batch of small orders to pick on a picking tour of the forward picking area. Each cart is equipped with fixed or moveable dividers or totes to allow the order picker to sort or pack merchandise into orders as he traverses the picking area. Pick lists define each picking tour by listing in location sequence the items and quantities on each picking tour. The

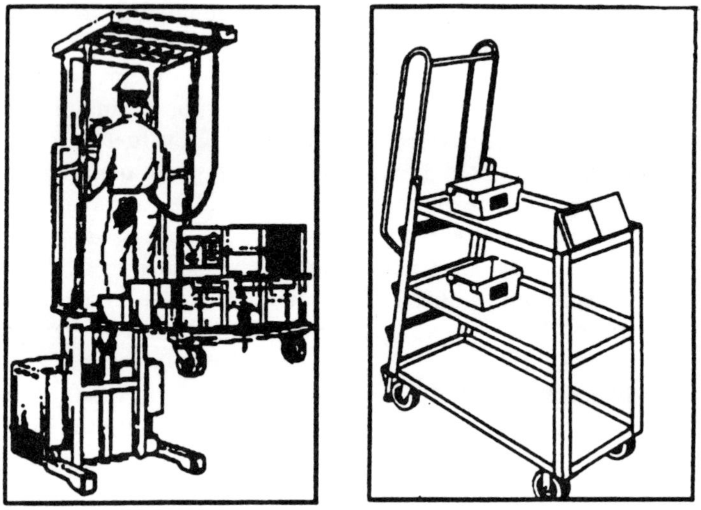

(A) (B)

Figure 5.2 Typical vehicle for reserve picking and forward replenishment. (B) Typicall order picking cart for the forward area.

order associated with each item is also listed next to the item on the picking list to permit accurate sorting into compartments on the picking cart. By batching small orders into picking tours, the overall distance traveled between each pick is reduced.

2.1 Costs

For the purposes of our study, the cost of a cart picking system is the sum of the amortized cost of racks and shelving used to house the space allocated to the items in the forward area, and the labor cost due to picking and replenishment.

Capital cost

We express the cost of racks and shelving as $C_{\mathrm{Cap}} = C_{\mathrm{Equip}} V$, where V is the cubic feet of available storage and C_{Equip} is the amortized cost per cubic foot of storage equipment. We assume the depth d, width w, and height h of the storage unit to have been predetermined.

Labor cost

The cost due to picking and replenishment is proportional (through the average wage rate) to the total amount of time required to execute the picking and replenishment activities.

Cost of replenishment The replenishment activity proceeds in two steps: the replenishment quantities of each item in a batch of predetermined size B_1 are first picked from the reserve area, and then restocked in the forward area. The amount of time to pick the replenishment batch is $t_R = B_1 t_e + t_w^r$, where t_e and t_w^r are respectively the time it takes to extract an item in the batch and t_w^r is the travel time to complete the batch. Since the layout of the reserve area is known, t_R is a known constant.

With regards to estimating the time to restock the replenishment batch, we assume that locations of items in the forward area are uniformly distributed. Consequently, the expected travel distance is $k_1\sqrt{AB_1}$, where A is the square footage of the forward area and k_1 is a proportionality constant typically in the range of 0.6 to 0.9 (see Daganzo 1984a,b). Letting θ denote the labor cost per hour and k_2 denote the proportionality constant converting distance into time, the cost of a single replenishment batch is

$$\theta\left(t_R + B_1 t_e + k_1 k_2 \sqrt{B_1 V/\rho}\right), \tag{1}$$

where the area A has been expressed in terms of the volume V through the density factor $\rho = 2dh/(2d+a)$ where a = aisle width. As a function of the set of items assigned to the forward area, and the amount of space allocated to these items, the total number of replenishment batches is

$$\frac{\Sigma_{i=1}^{N} x_i D_i / v_i}{B_1}, \tag{2}$$

where N is the number of items in the warehouse; x_i is a binary variable which is 1 if item i is assigned to the forward area, and 0 if it is not; v_i is the allocation in cubic feet to item i in the forward area; and D_i is the annual demand for item i measured in cubic feet. An obvious modification to (2) would model safety stocks, but to ease notational burden we have omitted explicit representation. As a function of the decision variables x_i, v_i, $i = 1,2, \ldots N$, and the size V of the forward area, the total replenishment cost, C_{Repl}, is

$$C_{\text{Repl}} = \theta\left(t_R + B_1 t_e + k_1 k_2 \sqrt{B_1 V/\rho}\right) \times \frac{\Sigma_{i=1}^{N} x_i D_i / v_i}{B_1}. \tag{3}$$

Cost of order picking Let B_2 and B_3 denote the predetermined batch size for picking items in the reserve and forward areas, respectively. The cost for picking items not assigned to the forward area, C_{PickR}, is

$$C_{\text{PickR}} = \theta\left(\frac{\Sigma_{i=1}^{N}(1-x_i)R_i}{B_2}\left(t_w^p + B_2 t_e\right)\right), \tag{4}$$

where R_i denotes the annual number of requests for item i, and t_w^p denotes the travel time to complete a batch. Following the logic used to estimate the cost to restock a replenishment batch, the cost for picking items assigned to the forward area, C_{PickF}, is

$$C_{\mathrm{PickF}} = \theta\left(B_3 t_e + k_1 k_2 \sqrt{B_3 V/\rho}\,\right) \times \frac{\sum_{i=1}^{N} x_i R_i}{B_3}. \tag{5}$$

The total cost, C_{Total}, of capital and labor as a function of the assignment vector x, allocation vector v, and size V of the forward area is $C_{\mathrm{Total}} = C_{\mathrm{Cap}} + C_{\mathrm{PickF}} + C_{\mathrm{Repl}} + C_{\mathrm{PickR}}$.

2.2 The constraints

While minimizing total cost is a goal the assignments and allocations chosen must be consistent with the available storage capacity:

$$\sum_{i=1}^{N} x_i v_i \leqslant V. \tag{6}$$

The assignments and allocations must also be chosen to avoid *congestion*. Congestion is defined to occur when the number of workers per square foot at an instant in time exceeds a threshold ω. As an approximation we insist that the average number of workers per square foot is bounded above by ω, namely,

$$\frac{T_{\mathrm{F}}}{\tau A_{\mathrm{w}}} \leqslant \omega, \tag{7}$$

where T_{F} denotes the annual labor time due to picking and replenishment in the forward area, τ denotes the annual number of warehouse hours dedicated to staffing the picking and replenishment activities in the forward area, and A_{w} denotes the walking area in the forward area. In terms of our parameters and decision variables

$$T_{\mathrm{F}} = \left(t_e + k_1 k_2 \sqrt{\frac{V}{\rho B_1}}\right) \sum_{i=1}^{N} x_i \frac{D_i}{v_i} + \left(t_e + k_1 k_2 \sqrt{\frac{V}{\rho B_3}}\right) \sum_{i=1}^{N} x_i R_i, \tag{8}$$

$$A_{\mathrm{w}} = V\left(\frac{1}{\rho} - \frac{1}{h}\right). \tag{9}$$

Congestion may not be a problem in cart picking systems if aisle widths easily allow more than one order picker to work in the same aisle. Though modeled differently, congestion is a common problem in end-of-aisle picking systems such as carousels and the miniload automated storage/retrieval system, where throughput may be constrained by the velocity of the storage/retrieval devices. In such systems congestion occurs when the total number of picks and internal replenishments exceed a threshold.

2.3 The mathematical model

The forward–reserve problem may be described as follows: Determine the forward-storage volume V, assignment and allocation vectors x and v, respectively, to minimize C_{Total} subject to the storage-volume capacity and congestion constraints, equations (6) and (7). From the mathematical perspective the forward–reserve problem may be expressed as the following optimization model:

$$\text{(FRP):} \qquad \theta\left(t_e + \frac{t_w^p}{B_2}\right)\sum_{i=1}^{N} R_i + \min_{V}\{C_{\text{Equip}}V - L(V)\}, \tag{10}$$

where $L(V)$ is the solution to the following 'Assignment–Allocation' (AA) optimization subproblem:

$$\text{(AA)} \qquad L(V) = \max_{x,v}\Sigma_{i=1}^{N} x_i\left(aR_i - b\frac{D_i}{v_i}\right) \tag{11}$$

subject to:

$$\Sigma_{i=1}^{N} x_i v_i \leqslant V, \tag{12}$$

$$\Sigma_{i=1}^{N} x_i\left(\alpha R_i + \beta\frac{D_i}{v_i}\right) \leqslant V, \tag{13}$$

$$v_i \geqslant 0,\ x_i \in \{0,1\},\ i = 1,2,\ldots,N, \tag{14}$$

and where

$$a = \theta\left(\frac{t_w^p}{B_2} - k_1k_2\sqrt{\frac{V}{\rho B_3}}\right),\ b = \theta\left(\frac{t_R}{B_1} + t_e + k_1k_2\sqrt{\frac{V}{\rho B_1}}\right),$$

$$\alpha = \frac{t_e + k_1k_2\sqrt{\dfrac{V}{\rho B_3}}}{\omega\tau(1/\rho - 1/h)}, \qquad \beta = \frac{t_e + k_1k_2\sqrt{\dfrac{V}{\rho B_1}}}{\omega\tau(1/\rho - 1/h)}. \tag{15}$$

For notational convenience in the sections to follow we have suppressed the functional dependence of a, b, α and β on V.

Remark 1

Hackman and Rosenblatt (1990) provide a heuristic solution to problem (AA) without the congestion constraint, and without parameters being dependent on the storage volume V. Hackman and Platzman (1990) develop a mathematical programming procedure that solves a generalization of the Hackman–Rosenblatt model to accommodate multiple forward-picking areas.

3 SOLUTION PROCEDURE

The formal description of our algorithm is the subject of this section. We begin by revealing the requisite mathematical structure embedded in (FRP), which we shall exploit to develop our algorithms.

3.1 Mathematical properties of program (FRP)

For each assignment vector x let

$$S(x)=\left\{v:\ v\geqslant 0,\ \sum_{i=1}^{N} x_i v_i \leqslant V,\ \sum_{i=1}^{N} x_i\left(\alpha R_i+\beta\frac{D_i}{v_i}\right)\leqslant V\right\}$$

and define

$$X=\{x:\ x\in\{0,1\} \text{ and } S(x)\neq\varnothing\}. \tag{16}$$

With this notation (AA) may be expressed as

$$\max_{x\in X}\ \max_{v\in S(x)}\ F(x,v) \tag{17}$$

where $F(x,v)$ denotes the objective function of (AA). After several algebraic manipulations it is possible to express and therefore replace the congestion constraint, equation (13), with

$$\left(\frac{\alpha b+a\beta}{\beta}\right)\Sigma_{i=1}^{N} x_i R_i - V\frac{b}{\beta}\leqslant F(x,v). \tag{18}$$

Observe that for $x\in X$ equation (18) will always be satisfied at the optimum $v^*(x)$. Therefore, for $x\in X$ it is possible to consider only the *inner maximization* (AA_x) defined as

$$(\text{AA}_x):\qquad \max_{v}\ F(x,v)$$
$$\text{subject to: } \Sigma_{i=1}^{N} x_i v_i\leqslant V. \tag{19}$$

The optimal solution to (AA_x) (Hackman and Rosenblatt, 1990) is

$$v_i^*(x)=\frac{\sqrt{D_i}}{\Sigma_{j=1}^{N} x_j\sqrt{D_j}}V. \tag{20}$$

After substituting $v^*(x)$, equation (20), into (18), and deleting the constant term $\theta(t_e+t_w^P/B_2)$ from equation (10), we reformulate (FRP) as

$$(\text{FRP})\qquad \min_{x,V}\left\{C_{\text{Equip}}\cdot V-\left(a\Sigma_{i=1}^{N} x_i R_i\right)+\frac{b}{V}\left(\Sigma_{i=1}^{N}\sqrt{D_i}x_i\right)^2\right\} \tag{21}$$

$$\text{subject to:} \quad \alpha \Sigma_{i=1}^{N} R_i x_i + \frac{\beta}{V}\left(\Sigma_{i=1}^{N} \sqrt{D_i} x_i\right)^2 \leqslant V \tag{21b}$$

$$x_i \in \{0,1\},\ i=1,\ 2,\ \ldots,\ N. \tag{21c}$$

Let $(\overline{\text{FRP}})$ be the program obtained by relaxing the binary constraint, equation (21c), with $0 \leqslant x_i \leqslant 1$, $i=1, 2 \ldots, N$. In Theorem 1 below we assume that the items have been relabeled so that

$$\frac{R_i}{\sqrt{D_i}} > \frac{R_{i+1}}{\sqrt{D_{i+1}}},\ i=1,\ 2,\ \ldots,\ N. \tag{22}$$

For convenience we assume no ties. Theorem 1 shows that the ratio defined in (22) determines a natural ordering of the items with respect to their desirability for assignment to the forward area. Accordingly we shall call this ratio the economic assignment quotient (EAQ).

Theorem 1. *Let x^* be an optimal solution to $(\overline{\text{FRP}})$. If $x_j^* > 0$ then $x_i^* = 1$ for $i<j$. If $0 < x_j^* < 1$ then $x_i^* = 0$ for $i>j$.*

Proof. To establish the claim we evaluate $\dfrac{\partial \mathfrak{L}(x,V,\eta)}{\partial x_i}$ of the Lagrangian associated with $(\overline{\text{FRP}})$ and exploit the well-known fact (Bazaraa and Shetty, 1979) that a necessary condition for a minimum of the Lagrangian with a lower- (respectively upper-) bound constraint is respectively $\partial \mathfrak{L}/\partial x_i \geqslant 0$ $(\partial \mathfrak{L}/\partial x_i \leqslant 0)$. Letting η denote the Langrange multiplier associated with the quadratic constraint, equation (21b), the Lagrangian is

$$\mathfrak{L}(x,V,\eta) = (C_{\text{Equip}} - \eta)V - (a - \eta\alpha)\sum_{i=1}^{N} R_i x_i + \frac{b+\eta\beta}{V}\left(\sum_{i=1}^{N} \sqrt{D_i} x_i\right)^2. \tag{23}$$

From equation (23)

$$\frac{\partial \mathfrak{L}(x,V,\eta)}{\partial x_i} = -(a-\eta\alpha)R_i + 2\sqrt{D_i}\left(\sum_{i=1}^{N} \sqrt{D_i} x_i\right)^2 \frac{\beta\eta + b}{V}. \tag{24}$$

Let (x^*, V^*) be the optimal solution to $(\overline{\text{FRP}})$ and let η^* denote the corresponding optimal multiplier. If $x_j^* > 0$ then $\dfrac{\partial \mathfrak{L}(x^*,V^*,\eta^*)}{\partial x_j} \leqslant 0$, and so from equation (24)

$$\frac{R_j}{\sqrt{D_j}} \geqslant 2\left(\sum_{i=1}^{N} \sqrt{D_i} x_i^*\right)^2 \frac{b+\beta\eta^*}{V(a-\eta^*\alpha)}. \tag{25}$$

Since $\dfrac{R_i}{\sqrt{D_i}} > \dfrac{R_j}{\sqrt{D_j}}$ for all $i<j$ it follows directly from equation (25) that

$\frac{\partial \mathcal{L}(x^*,V^*,\eta^*)}{\partial x_i}<0$, and so $x_i^*=1$. If $0<x_j^*<1$ then equation (25) also shows that $\frac{\partial \mathcal{L}(x^*,V^*,\eta^*)}{\partial x_i}>0$ for $i>j$ and so $x_i^*=0$. The result now follows.

Remark 2

An inspection of equation (25) shows that the EAQ, equation (22), is *independent* of V.

The algorithms described in the next section are based on the EAQ exhibited in Theorem 1 and the unimodality of the functions ν_k and r_k from $\{1,2,. . . ,N\}\mapsto R_+$ defined by

$$\nu_k = a\sum_{i=1}^{k} R_i - b\frac{\left(\sum_{j=1}^{k}\sqrt{D_j}\right)^2}{V} \tag{26}$$

and

$$r_k = \alpha\sum_{i=1}^{k} R_i + \beta\frac{\left(\sum_{j=1}^{k}\sqrt{D_j}\right)^2}{V} \tag{27}$$

The proof that ν_k is unimodal can be found in Hackman and Rosenblatt (1990), while r_k is unimodal by definition. Note that for

$$x = [\overbrace{1,1,. . . ,1}^{k},\overbrace{0,0,. . . ,0}^{N-k}],$$

$\nu_k = F(x,\upsilon^*(x))$, equation (18), while r_k measures congestion, equation (21b).

3.2 Algorithm description and validation

In what follows we will suppress the dependence of the program on V. Nevertheless the conclusions hold for each feasible V.

Let (x^*,υ^*) denote the optimal solution to $(\overline{\text{FRP}})$, let $K^*=\{i: x_i^*>0\}$ denote the optimal set of items assigned to the forward area. Originally, for a given volume V, K^* could have been any one of the 2^N possible assignment vectors. However, as a consequence of Theorem 1 the set K^* must be equal to $\{1,2,. . . ,k^*\}$ for some $1\leqslant k^*\leqslant N$. Consequently, there are now only N possibilities. It follows further from Theorem 1 that the solution to $(\overline{\text{FRP}})$ may have *at most* one component with a fractional value, i.e. $0<x^{k^*}<1$ which necessarily is the 'last' item assigned to the forward area. This will happen if equation (25) holds as an equality. Since the number of items N in practice is of the order of tens of thousands, a *near-optimal* solution to $(\overline{\text{FRP}})$ can be found by finding k^*, and simply ignoring the assignment of item k^* to the forward area. Let x^{k^*} denote the near-optimal assignment vector associated with k^*, i.e. $x_i^{k^*}=1$ if $i<k^*$, and $x_i^{k^*}=0$ if $i\geqslant k^*$. Since the set

$$S(x^{k^*})=\left\{v\geqslant 0:\ \sum_{i=1}^{(k^*-1)} v_i\leqslant V,\ F(x^{k^*},v)\geqslant\frac{(b\alpha+\beta a)\Sigma_{i=1}^{(k^*-1)} R_i-bV}{\beta}\right\} \tag{28}$$

is nonempty when maximizing $F(x^{k^*},v)$, the congestion constraint, equation (21b), may be ignored as before. The solution v^{k^*} is given, equation (20), by

$$v_i(x^{k^*})=\frac{\sqrt{D_i}}{\Sigma_{j=1}^{(k^*-1)}\sqrt{D_j}}V,\ i=1,2,\ldots,(k^*-1). \tag{29}$$

One can obtain k^* by direct enumeration, namely, by evaluating sequentially for $k=1,2,\ldots$ its corresponding performance, ν_k, given by equation (26), while ensuring that the congestion constraint, equation (13), is not violated by evaluating r_k, equation (27). Consequently, a *direct enumeration algorithm* can be used to solve (FRP). Such enumeration produces a feasible solution which differs from ($\overline{\text{FRP}}$) by at most one item, and since the optimal objective function for the relaxation ($\overline{\text{FRP}}$) cannot be higher than the optimal objective function value for (FRP), the direct enumeration quickly produces a near-optimal solution to (FRP), as desired.

The volume, V, of the forward area is directly proportional to the number of shelving equipment units. The number of feasible shelving equipment units is of the order of a few hundreds, and so the solution of the FRP is obtained by determining the optimal assignment–allocation, $k^*(V)$, for each $V\in \boldsymbol{V}$ is the set of feasible alternatives, $\boldsymbol{V}=\{V_1,V_2,\ldots,V_S\}$.

Algorithm I

The algorithm consists of two phases. In Phase I, the optimal assignment and optimal volume for the forward area are determined. In Phase II, the optimal volume of the individual items, v_i, is calculated.

Phase I

Step (0): Relabel the items so that

$$\frac{R_i}{\sqrt{D_i}}>\frac{R_{i+1}}{\sqrt{D_{i+1}}},\ i=1,2,\ldots,N.$$

Initialization: Set C to $+\infty$ and the two-dimensional vector $[p_1,p_2]$ to zero.

Step (1): For $\ell=1,2,\ldots,S$ begin:

Step (2): For $k=1,2,\ldots$ compute sequentially

$$\nu_k^\ell = a\sum_{i=1}^{k} R_i - b\frac{\left(\sum_{j=1}^{k}\sqrt{D_j}\right)^2}{V_\ell}, \text{ and}$$

$$r_k^\ell = \alpha\sum_{i=1}^{k} R_i + \beta\frac{\left(\sum_{j=1}^{k}\sqrt{D_j}\right)^2}{V_\ell}.$$

Step (3): Stop whenever $r_{k+1}^\ell > V_\ell$ or $\nu_k^\ell > \nu_{k+1}^\ell$.

Step (4): If $(C_{\text{Equip}}\cdot V_\ell - \nu_k^\ell) < C$, set $C=(C_{\text{Equip}}\cdot V_\ell - \nu_k^\ell)$ and update $[p_1,p_2] = [\ell,k]$. If $\ell<S$ set $\ell=\ell+1$ and go to Step (2). Else continue.

Let $[p_1,p_2] = [\ell^*,k^*]$. Define $c=\dfrac{R_{k^*}}{\sqrt{D_{k^*}}}$ and $\gamma=\dfrac{\sum_{j=1}^{k^*}\sqrt{D_j}}{N}$ as control parameters.

Phase II

Step (5): Assign each item with $\dfrac{R_i}{\sqrt{D_i}} \geqslant c$ to the forward area, allocating to item i

$$v_i = \frac{\sqrt{D_i}V_{\ell^*}}{\gamma N}$$

units of volume.

When the set of possible storage volumes is extremely large, it may be more convenient to view the set V as continuous. In this case it is appropriate to reverse the order of decision-making by first determining the optimal volume, $V^*(k)$, for each *ratio-ordered* feasible assignment–allocation, k. This is the basis for our second algorithm below.

Algorithm II

Here we can no longer suppress the dependence of the parameters a, b, α, and β, equation (15), on the volume V. Each of these parameters is defined by a sum or a difference of two terms, one independent of V and the other proportional to $\sqrt{V}$. We respectively denote the corresponding coefficients as a_i, b_i, α_i, and β_i, $i=1,2$, and express the parameters, equation (15), as

$$a = a_1 - a_2\sqrt{V},\ b = b_1 + b_2\sqrt{V},$$

$$\alpha = \alpha_1 + \alpha_2\sqrt{V},\ \beta = \beta_1 + \beta_2\sqrt{V}.$$

After deleting the term $-a_1\Sigma_{i=1}^{N} x_i R_i$, which does not depend on V, the problem (FRP) now becomes

$$(\mathrm{FRP}_k) \qquad \min_{V \in V} a_2\left(\sum_{i=1}^{k} R_i\right)\sqrt{V} + b_1\left(\sum_{i=1}^{k}\sqrt{D_i}\right)^2 V^{-1} + b_2\left(\sum_{i=1}^{k}\sqrt{D_i}\right)^2 V^{1/2}$$

Subject to:

$$\alpha_1\left(\sum_{i=1}^{k} R_i\right)V^{-1} + \alpha_2\left(\sum_{i=1}^{k} R_i\right)V^{-1/2} + \beta_1\left(\sum_{i=1}^{k}\sqrt{D_i}\right)^2 V^{-2} + \beta_2\left(\sum_{i=1}^{k}\sqrt{D_i}\right)^2 V^{-3/2} \leqslant 1,$$

where the notation (FRP_k) emphasizes the dependence of the solution on the choice of k. This program is a geometric program (Duffin *et al.*, 1967) and as such is *convex transformable*. Consequently, every stationary point is a global minimum. Moreover, the polynomial (Duffin *et al.*, 1967) constraint when viewed as an equality has only one *positive* root, x_R. Thus, the minimum can be determined by a bisection search over the semi-open interval $[x_R, +\infty)$. It is unnecessary to explictly calculate x_R, since $x < x_R$ if and only if the above constraint is violated.

Phase I

Let $G_k(V)$ denote the objective function. The zeroth and initialization steps are the same for both methods. We will start therefore from Step (1).

Step (1): For $k = 1, 2, \ldots, N$ begin:

Step (2): Solve (FRP_k) to determine V_k^*.

Step (3): If $(G_k(V_k^*) - a_1\Sigma_{i=1}^{k} R_i) < C$, set $C = (G_k(V_k^*) - a_1\Sigma_{i=1}^{k} R_i)$ and update $[p_1, p_2] = [\ell, k]$. If $k < N$ set $k = k + 1$ and go to Step (2). Else continue.

The remaining steps are the same for both methods.

4 A STATISTICAL PROCEDURE TO SOLVE THE FORWARD–RESERVE PROBLEM

Conceptually Algorithm I presented in Section 3 consists of two phases. Phase (I) determines two *control* parameters c and γ. Phase II assigns an item to the forward area if its corresponding ratio $R/\sqrt{D}$ exceeds c; if the item is assigned to the forward area, its allocation is $\dfrac{\sqrt{D}V_\ell}{\gamma N}$. (In the context of Section 3, $c = R_{k^*}/\sqrt{D_{k^*}}$.) Viewed from this perspective, accurate assignment–allocations can be obtained by solving Phase I using only a small number of items. First, select $n \ll N$ items at random. Next, the available forward volume is proportionally adjusted by

$V_\ell^n \equiv (n/N)V_\ell$. Solve Phase I of the reduced FRP to determine the control parameters c^n and γ^n. Then, proceed with Phase II but now with the *entire* population, see Step (5). The function $F(x,v)$, equation (18), measures the savings in operating cost due to the introduction of a forward storage area. The term $L(V)$, equation (11), equals the optimum savings. $F(x,v)$ is the sum of the savings for each individual item in the population. We let f^n denote the average overall savings per item determined by this procedure. For the procedure to work the control parameters c^n and γ^n determined from the sample of items must be sufficiently close to control parameters c^N and γ^N determined from solving the problem with the entire population. Under a reasonable statistical model we will show that the sequences $\{c^n\}$, $\{\gamma^n\}$, and $\{f^n\}$ converge. The implication is that for large N and for n sufficiently large, $c^n \approx c^N$, $\gamma^n \approx \gamma^N$, and $f^n \approx f^N$, as required. Standard statistical practice suggests that only a few hundred sample items are needed. Consequently, if (R,D) tends to follow some standard joint distribution, tables listing c and γ in terms of simple parameters like means and variances may be estimated from sample data. Then, given a system configuration, the assignment and allocation subproblem could be solved on a calculator!

Furthermore, Phases I and II may be decoupled. Phase I is performed periodically, say, twice a year, to allow the parameters c and γ to respond to shifts in demand and introduction/retirement of stock items. Phase II is performed on an individual stock item whenever it must be handled in bulk, when new stock arrives and must be stored, or when forward stock is exhausted and must be replenished. The following theorem is proved in an appendix.

Theorem 2 *The sequences $\{c^n\}$, $\{\gamma^n\}$, and $\{f^n\}$ converge almost surely (a.s.).*

5 CASE STUDY

The following case study is based on data provided by a major distributor/retailer of casual apparel, sporting goods, and wares for home and camp use. (Henceforth we refer to the company as XYZ, Inc.) Detailed data specifying each of the parameters defined in our forward–reserve model was not available. We were able to obtain approximate time-study measurements of warehouse activities, aggregate parameters describing the activity distribution of the items in the warehouse, and sketch of the warehouse layout. The data is sufficient to conduct a forward–reserve analysis, which will illustrate the affect of properly sizing the forward picking area by using the EAQ to select the right items in their right quantities.

At XYZ, Inc. approximately \$500 million in annual sales are generated from 12 million orders for 25 million units. The company is experiencing 25% annual growth in sales and 5% annual growth in the number of items on-hand.

XYZ, Inc. houses more than 50 000 unique items in a 450 000 square-foot distribution center. 160 000 square feet are allocated to reserve storage. The storage and retrieval of cases in the reserve area is automated in several narrow aisles. The forward picking area consists of 100 aisles of bin shelving and flow rack in 150 000 square feet of space. The forward area can house 200 000 cubic feet of storage.

Each of the 50 000 items receives a location in the forward area. For each item, space is provided to house 20 units. A random storage-assignment policy is employed in both the forward and reserve areas. Random storage is employed in an effort to conserve space in the reserve area and to minimize congestion in the forward area.

A typical customer order is for 2 line items; 85% of all picks can be completed with 15% of the items. During each picking tour, order pickers complete between 20 and 30 orders. Orders are picked into pigeon-hole carts that allow the pickers to sort while they are picking. The travel time per line-item pick in the forward area requires approximately ½ minute. The travel time per line-item pick from the reserve area requires approximately 2 minutes. The average time to complete a line-item replenishment from the reserve area to the forward area requires approximately 4 minutes. Labor costs including fringes are approximately $30/hour per employee.

We first consider the annual labor cost due to picking and replenishment in the present system. Each line-item pick in the forward area costs $0.25. Since all picks are executed in the forward area the total annual picking cost due to labor is $6.25 million. The average annual demand per item is 500 units. Since each item is allocated 20 units in the forward area each item requires 25 replenishments per year on average. Thus, the total number of line-item replenishments per year is 1.25 million. At $1.00 per replenishment, the annual replenishment cost is $1.25 million. Thus the total annual labor cost due to picking and replenishment is $7.5 million.

With 25 million picks per year there is a considerable opportunity for cost reductions by reducing the size of the forward picking area. Also, since the aisle widths easily permit multiple pickers in the same aisle, congestion is not an issue.

Following the procedure outlined in Section 3 we first ranked the items according to the ratio of the annual number of picks to the square root of the annual demand measured in cubic feet. We then iterated over the number of aisles dedicated to the forward area to determine a more cost-effective forward-area size. We found that the number of aisles should be closer to 32, and the floorspace reduced to 50 000 square feet, 32% of its original size.

By reducing the size of the forward area to 32% of its original size the travel cost to pick in the forward area has been lowered from $0.25 to $0.14. (The square root of 0.32 is 0.58 and $0.58 \times 0.25 = 0.14$.) The solution is to allocate space in the forward area for roughly 50% of the items—representing 22 million picks or 88% of the picking activity. The new annual labor cost for picking in the forward area would be $3.0 million. To pick the remaining 3 million picks in the reserve area would cost $1.5 million, and so the new cost of replenishment would be $1.5 million. In sum, the total annual labor cost due to picking and replenishment associated with the proposed solution generated from our algorithm of Section 3 would be $6.0 million, an annual labor-cost saving of $1.5 million. Note that this excludes the savings in equipment operating costs due to a smaller forward area.

Finally, we note that even if each item must be allocated some space in the forward area, and all 150 000 square feet must be used for order picking, solution of the assignment–allocation (AA) subproblem would reallocate the space to significantly reduce the annual replenishment cost. For XYZ, Inc., the total annual number of replenishments would be reduced from 1.25 million to 0.95 million, a reduction of almost 25%. The associated annual labor savings would be $300 000.

6 EXTENSIONS

There are a variety of order-picking systems in which near-optimal solutions to the forward–reserve problem can increase order-picking productivity. For example, a popular method for case picking is pallet jack picking from the floor level of 4- or 5-high pallet rack. The floor-level locations are dedicated case-picking locations and the upper-level locations are random for reserve storage and pallet picking. In this scenario, the floor-level case-picking locations define the forward picking area. The solution procedure defined here is directly applicable, with restocking transactions defined as the retrieval of a pallet from a random-reserve location and storage in a dedicated floor-level case-picking location.

Other examples arise in the configuration of automated order-picking systems. Since the forward picking area might be a horizontal or a vertical carrousel, a miniload automated storage/retrieval machine, or an automated dispensing machine, the cost of each cubic foot of forward picking space can range from $30 to $1000. Hence, in automated order-picking systems, the determination of the optimal forward picking allocation has a major impact on capital expense as well as on overall operating productivity.

APPENDIX THE PROOF OF THEOREM 2

Theorem 2 is based on the following statistical model. We view R and D as possibly nonindependent random variables, following a smooth (although possibly unknown) distribution $g(R,D)$. We consider a sequence of problems (C^n) parametrized by n, where the nth problem is based on the first n items of the sequence. We assume the density g is zero when R or D is nonpositive, and positive when both R and D are positive. Our analysis invokes the law of large numbers, so we assume the expectation of R^2+D^2 to be finite.

We shall prove the theorem by establishing a sequence of lemmas. Let $I(A)$ denote the indicator of the event A. Define the following sequences of random variables:

$$\mathfrak{R}^n(c)=\frac{\sum_{i=1}^{n} R_i\cdot I\left(\frac{R_i}{\sqrt{D_i}}>c\right)}{n}, n=1,2,\ldots,$$

$$\mathfrak{D}^n(c)=\frac{\sum_{i=1}^{n} \sqrt{D_i}\cdot I\left(\frac{R_i}{\sqrt{D_i}}>c\right)}{n}, n=1,2,\ldots,$$

$$f^n(c)=a\mathfrak{R}^n(c)-\frac{b(\mathfrak{D}^n(c))^2}{V_\ell^n/n},\ n=1,2,\ldots,$$

$$p^n(c)=\alpha\mathfrak{R}^n(c)+\frac{\beta(\mathfrak{D}^n(c))^2}{V_\ell^n/n},\ n=1,2,\ldots,$$

By the law of large numbers,

$$\mathfrak{R}^n(c)\to\mathfrak{R}^\infty(c)\equiv \boldsymbol{E}\left\{R\cdot I\left(\frac{R}{\sqrt{D}}>c\right)\right\}\text{a.s.,} \tag{A1}$$

$$\mathfrak{D}^n(c)\to\mathfrak{D}^\infty(c)\equiv \boldsymbol{E}\left\{\sqrt{D}\cdot I\left(\frac{R}{\sqrt{D}}>c\right)\right\}\text{a.s.} \tag{A2}$$

Further define

$$f^\infty(c)=a\mathfrak{R}^\infty(c)-\frac{b(\mathfrak{D}^\infty(c))^2}{V_\ell/N},$$

$$p^\infty(c)=\alpha\mathfrak{R}^\infty(c)+\frac{\beta(\mathfrak{D}^\infty(c))^2}{V_\ell/N},$$

and observe that

$$V_\ell^n/n=V_\ell/N.$$

From equations (A1) and (A2), each sequence of random variables $\{f^n(c)\}$ and $\{p^n(c)\}$ converges to $f^\infty(c)$ and $p^\infty(c)$ a.s., respectively. Since $p^\infty(\cdot)$ is continuous and strictly decreasing there exists a unique constant c_p^∞ for which

$$p^\infty(c_p^\infty)=V_\ell/N.$$

Since $\mathfrak{D}^\infty(\cdot)$ is continuous and strictly decreasing there exists a unique c_f^∞ for which

$$\mathfrak{D}^\infty(c_f^\infty)=\frac{ac_f^\infty\cdot V_\ell/N}{2b}.$$

Lemma 1. *$f^\infty(\cdot)$ is strictly increasing on $(0,c_f^\infty]$ and strictly decreasing on $[c_f^\infty,\infty)$.*

Proof. From calculus

$$\frac{\mathrm{d}f^\infty(c)}{\mathrm{d}c}=\left[\int_0^\infty\sqrt{D}g(c\sqrt{D},D)\mathrm{d}D\right]\left(\frac{2b}{V_\ell/N}\mathfrak{D}^\infty(c)-ac\right).$$

Since g is positive on the domain of integration, the derivative of $f^\infty(c)$ with respect to c is positive on $(0,c_f^\infty)$ and negative on (c_f^∞,∞), and so the result follows.

Lemma 2. *The sequence of random variables $\{c^n\}$ converges to $c^\infty \equiv \max(c_f^\infty, c_p^\infty)$ a.s.*

Proof. $f^n(\cdot)$ is piecewise-constant and unimodal (Hackman and Rosenblatt, 1990). Consequently, it is possible to define the smallest random variable c_f^n which maximizes $f^n(\cdot)$. $p^n(\cdot)$ is piecewise-constant, right-continuous, and nondecreasing. Consequently, it is possible to define the random variable c_p^n that records the smallest constant c_0 for which $p^n(c_0) \leqslant V_\ell / N$. In terms of c_f^n and c_p^n,

$$c^n = \max(c_f^n,\ c_p^n). \tag{A3}$$

Pick two positive numbers c_1 and c_2 for which $c_1 < c^\infty < c_2$. Since f^n converges to f^∞ a.s., eventually, $f^n(c_1) \approx f^\infty(c_1)$, $f^n(c_2) \approx f^\infty(c_2)$ and $f^n(c^\infty) \approx f^\infty(c^\infty)$. Lemma 1 and the unimodality of f^n ensure that eventually $c^n \in [c_1, c_2]$. Since c_1 and c_2 were chosen arbitrarily, c_f^n converges to c^∞ a.s. Since p^∞ is strictly decreasing and p^n is nonincreasing, a similar analysis shows that c_p^n converges to c_p^∞ a.s. The result now follows from equation (A3).

Lemma 3. $\mathfrak{D}^n(c^n) \to \mathfrak{D}^\infty(c^\infty)$ *and* $f^n(c^n) \to f^\infty(c^\infty)$ *a.s.*

Proof. Pick an $\epsilon > 0$, and pick two positive numbers c_1 and c_2 for which $c_1 < c^\infty < c_2$. Since $\mathfrak{D}^n$ is nonincreasing, and since $c^n \to c^\infty$ a.s. by Lemma 2, eventually, $\mathfrak{D}^n(c^n) \leqslant \mathfrak{D}^n(c_1) \leqslant \mathfrak{D}^\infty(c_1) + \epsilon$ and $\mathfrak{D}^n(c^n) \geqslant \mathfrak{D}^n(c_2) \geqslant \mathfrak{D}^\infty(c_2) - \epsilon$. Eventually, therefore,

$$\mathfrak{D}^\infty(c_2) - \epsilon \leqslant \mathfrak{D}^n(c^n) \leqslant \mathfrak{D}^\infty(c_1) + \epsilon. \tag{A4}$$

Now, $\mathfrak{D}^\infty$ is continuous and strictly decreasing; therefore, since ϵ was chosen arbitrarily, as c_1 and c_2 approach c^∞, equation (A4) shows that $\mathfrak{D}^n(c^n)$ will converge to $\mathfrak{D}^\infty(c^\infty)$ a.s. An analogous argument shows that $\mathfrak{R}^n$ converges to $\mathfrak{R}^\infty$. Consequently, $f^n(c^n)$ also converges to $f^\infty(c^\infty)$.

REFERENCES

Bazaraa, M. S., and Shetty, C. M. (1979) *Nonlinear Programming Theory and Algorithms*, John Wiley & Sons, New York.

Bozer, Y. A. (1985) *Optimizing Throughput Performance in Designing Order Picking Systems*, Ph.D. Dissertation, School of Industrial and Systems Engineering, Georgia Institute of Technology, November.

Daganzo, C. F. (1984a) The length of tours in zones of different shapes, *Transportation Research*, **18B**, 135–145.

Daganzo, C. F. (1984b) The distance travelled to visit *N* points with maximum of *C* stops per vehicle: An analytic model and an application, *Transportation Science*, **18**, 331–350.
Duffin, R. J., Peterson, E. L., and Zener, L. (1967) *Geometric Programming*, John Wiley & Sons, New York.
Hackman, S. T., and Platzman, L. K. (1990) Near optimal solution of generalized resource allocation problems with large capacities, *Operations Research*, **38** (5), 902–910.
Hackman, S. T., and Rosenblatt, M. J. (1990) Allocating items to an automated storage and retrieval system, *IIE Transactions*, **22** (1), 7–14.
WERC (1986) *Survey of Warehousing Professionals*, The Warehousing Education and Research Council, Chicago.
White, J. A., and Kinney, H. D. (1982) Storage and warehousing, in G. Salvendy (Ed.), *Handbook of Industrial Engineering*, John Wiley & Sons, New York.

School of Industrial and Systems Engineering
Georgia Institute of Technology
Atlanta
GA 30332
USA

6
Optimized Production Planning and Delivery Quotation for the Semiconductor Industry

Robert C. Leachman[1] and Dale J. Raar[2]
[1]*University of California at Berkeley, CA, USA*
[2]*Harris Corporation, Palm Bay, FL, USA*

ABSTRACT
Important aspects of a company-wide production-planning system implemented at Harris Corporation Semiconductor Sector are discussed. At the heart of the system is an optimization-based production-planning system known as the Berkeley Planning System (BPS), developed at the University of California at Berkeley. The system utilizes mathematical optimization techniques to establish worldwide factory production plans and product-availability schedules supporting the on-line quotation of delivery dates. This paper focuses on the choice of an objective function for planning decisions made by BPS. It is shown that overall demands for each product must be broken into classes such as on-hand customer orders, safety-stock rebuilds, reliable-demand forecasts, uncertain-demand forecasts, etc., in order to properly reflect business objectives. An approach somewhat similar to goal programming is utilized in which successive mathematical optimization calculations are performed to incrementally load the prioritized classes of demands on constrained production facilities using different objective functions for different demand classes.

1 BACKGROUND ON PRODUCTION PLANNING IN THE SEMICONDUCTOR INDUSTRY

Many approaches to the problem of producing a company-wide production plan have been put into practice around the world. In most semiconductor companies, an iterative, manually assisted approach has evolved internally within each company over a period of years. This approach generally involves a trial plan developed by a central planning department or planning staff within the company that is based

Optimization in Industry 2. Edited by T. A. Ciriani and R. C. Leachman

on customer-order and sales-forecast information obtained from the marketing and customer-service groups. This trial plan is sent out to the factory areas. At this point, factory-level planning staff scrutinize the plan and make changes to conform with capacity or raw-material limitations or with various manufacturing objectives such as linear loading, desirable lot sizes, etc. A revised plan is sent back from each factory to the central planning organization to be incorporated with all of the other factory plans. A final version of the worldwide production plan incorporating 'minor' modifications is then returned to the factories to be put into use, or perhaps more discussion and revision will ensue if more 'major' modifications have been made reflecting serious misgivings the marketing department may have had about the adjustments made by the factories. Last-minute changes because of overlooked data or new information may lead to more iterations. These iterations can take days to weeks to accomplish and usually are based on experience and 'gut feel' of planners.

Replacing the iteration among planning groups with an automated planning calculation is an important goal for several reasons. Stability and predictability of production planning would be enhanced. The number of personnel devoted to planning could be reduced. And, most importantly, the time to generate a plan could be significantly reduced (to a weekend or perhaps overnight). In many companies, the planning cycle now takes several weeks to a month. This means a major portion of the production plan is based on demand-forecast input rather than actual customer orders. If a plan could be generated overnight, the portion of the plan that is based on actual customer orders could become much higher, and this would result in an increase in the level of customer service which the production plan provides.

A major obstacle to systematizing these calculations is the decentralized nature of the planning process in almost every semiconductor company. If the initial production plan were feasible with respect to demands, work-in-process (WIP), raw material, and capacity, many if not all of the requirements for generating a company-wide plan that is acceptable by all involved would be realized. Additionally, rigorous procedures that rationalize and trade off the conflicting objectives of various organizations in an integrated calculation would be a major improvement over the current environment in which any group may propose changes which could be viewed unfavorably by other groups, leading to further discussion and delays in realizing a production plan.

2 THE BERKELEY PLANNING SYSTEM

The Berkeley Planning System (BPS), a production-planning system developed at the University of California, Berkeley, has been designed to automatically generate production plans for semiconductor manufacturers that are feasible with respect to WIP, raw material, and capacity limitations. BPS uses mathematical optimization techniques pursuing multiple objectives reflecting different business concerns as discussed below. The optimization models incorporated into the system are formally described in Leachman (1993).

BPS is a noniterative implementation. The computer-generated first-pass plans for end products meet all of the feasibility requirements discussed above and do not need to be further 'massaged'. This makes it possible to regenerate worldwide production plans on a much more timely basis than possible with manual, iterative procedures.

BPS has been installed and is currently being used by Harris Corporation Semiconductor Sector to generate production plans for five geographically separate manufacturing sites, with each site including multiple factories. The overall planning system, databases and methodology at Harris are known as IMPReSS (Integrated Manufacturing Production Requirements Scheduling System). The manufacturing organizations at various manufacturing sites of Harris have recently begun to use these plans as the basis for their operations with a minimal amount of 'judging' by planners. A new set of plans is calculated each week over the weekend and distributed to the manufacturing areas for their use by Monday morning local time. Previously, most of the Harris Semiconductor production plans were generated on a monthly basis with the final approved plan being completed during the second week of the month.

The goals of the Berkeley Planning System are: (a) to provide excellent on-time delivery to existing on-hand orders, and (b) to establish availability in response to forecast demands for the purposes of delivery quotation, subject to capacity and manufacturing-yield limitations. As will be discussed below, the plans generated by the BPS tend to meet demands as close to 'on-time' as possible.

BPS utilizes a rate-based model of production, i.e. factories are scheduled with constant production rates in user-defined planning periods such as weeks or months. BPS smooths uneven demand patterns with more practical production rates that meet the demands. The horizon condition for the production plan is to end the plan with steady-state production rates, rather than the usual termination of production condition that is more characteristic of closing down the factories. This condition is important in modeling semiconductor manufacturing because of the time-phased capacity utilization by reentrant manufacturing flows.

In order to cope with constraints concerning demand, capacity, and WIP in a single-pass production-planning system, the overall planning calculation within BPS is broken up into two basic stages—*requirements planning*, followed by *capacitated loading*. Optimization is applied at both of these stages of the calculation.

Optimization is necessary with respect to requirements planning to ensure that the complex product structure prevalent in the semiconductor-processing industry is handled correctly. Many semiconductor products are classified into various categories of performance quality (called binning) determined by a series of tests. Some of the product bins can be used to fill demands for other bins of the product which have compatible test requirements (called substitution). The combination of binning and substitution found here cannot be handled effectively using the standard materials requirements planning (MRP) planning methods. Instead, linear programming mathematical techniques are used to make the appropriate calculations and determine the 'optimal' net requirements. (Although many of the assumptions and decompositions required to make BPS tractable reduce the optimality of the production plan, the term 'optimal' will be used to describe the results of the component calculations in BPS-generated production plans.)

During the requirements-planning stage, new starts that will be needed are determined with as little overproduction as possible, taking into account current WIP projections, on-hand inventory, process characteristics, product structure, and rate-based production. Products which have a 'straight-through' structure are put into an MRP-type calculation. Those products which exhibit binning or substitution (or both) are handled with a linear programming (LP) approach.

Capacitated loading of the net requirements is done exclusively using linear programming optimization within BPS in order to maintain the feasibility of capacity and raw-material usage while meeting the dates of the net demand requirements as closely as possible. By incorporating both lateness costs and inventory-holding costs in the objective function, the goal of being just-in-time (JIT) toward meeting customer and other demands is a primary consideration for minimizing costs during this calculation. Available capacity will be used as much as is possible. However, capacity limitations may cause the plan to be 'late' for a demand. Excess capacity will go unused if there is no demand which would require its use.

Attempting to make good use of manufacturing assets is done while calculating the best dates that can be met. The total demand forecast (current orders, inventory rebuild, and forecast) is used as the major lever toward increasing capacity utilization and controlling costs. Capacity will be used toward building products which the company believes that it will be able to sell. In addition, those products which create a higher cash flow will get a higher priority for use of bottleneck capacity or scarce raw materials.

When it is not possible to service all demands on-time, it is necessary to account for the relative business importance of various demands and the relative certainty that positive cash flow from meeting the demands will actually materialize. Obviously, servicing previous customer commitments has a different kind of business importance than providing output meeting forecasts for future customer orders. Moreover, among demand forecasts, some may have much lower risk of not materializing then others, e.g. forecasts arising from customers counting on Harris as a long-term supplier for the customers' high-volume production lines versus forecasts arising from customers viewing Harris as a backup supplier in times of chip shortages, or customers that are new startups with grandiose plans for capturing market share.

3 SOFTWARE ARCHITECTURE OF BPS

BPS is partitioned into calculation *modules*, each performing either requirements planning or capacitated loading or both on a portion of the overall product structure (see Figure 6.1). Module 1 performs requirements planning on the final-test portion of the product structure, deriving net requirements for new final-test starts. This is an MRP-type calculation.

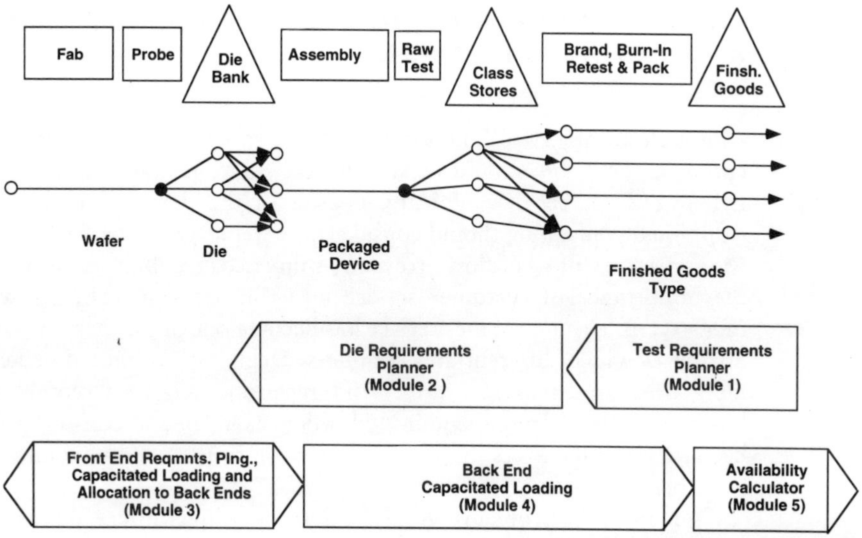

Figure 6.1 Software architecture of the IMPReSS planning engine (BPS).

Module 2 performs requirements planning on the initial test, assembly and die-bank portion of the product structure, to derive net requirements for new die output from wafer fab/wafer probe factories. Separate requirements-planning calculations are made for each product family; those families with straight-through (no binning or substitution) product structures are planned using MRP calculations, while those with binning and/or substitution are planned using linear programming optimization. Module 2 includes many relatively small optimization calculations.

Module 3 does requirements planning for new wafer output from fab and probe. This is followed by capacitated loading of these net requirements. Allocation of the planned die output among demands for new assembly starts (calculated in Module 2) is then performed. The requirements-planning portion involves either MRP calculations or linear programming optimization calculations for each product family (depending on its product structure). A single, large capacitated-loading calculation is performed by priority class for all wafer fabs and probes serving a particular product set. The allocation of planned output among assembly start demands is a network flow optimization model.

Module 4 performs capacitated loading of assembly and test, attempting to meet final output demands subject to die availability planned by Module 3, as well as capacity and raw-materials limitations. This is a very large optimization calculation, even when separate, parallel optimization calculations are defined for each assembly/test manufacturing site. At Harris, the largest Module 4 optimization model exceeds 100 000 rows.

At present, Harris Semiconductor has implemented BPS on IBM RISC System/6000 Model 560 workstations equipped with the Optimization Subroutine Library (OSL) software for solving mathematical optimization programs. At present, simplex-method mathematical optimization is used in Modules 1, 2, and 3, and interior-point methods are used to solve Module 4 problems.

4 THE CHALLENGE OF DEVELOPING A SUITABLE OBJECTIVE FUNCTION FOR MANUFACTURING PLANNING

Although an objective function is required in order to use linear programming, choosing what this should be is a difficult task. Minimizing cost of production will not work because typically not all of the demand forecast can be met. Additionally, planning should consider the revenue generated by a production plan. Maximization of cash flows (revenue minus cost) is better, but it does not take the importance of customer service levels into account. In the semiconductor industry, excellent customer service has become one of the most important business attributes which differentiates suppliers. Delaying committed orders in order to meet a forecast demand for more profitable products is not acceptable in the current business climate. Taking committed orders as a demand constraint will not work because it is sometimes infeasible to meet on-time all of the commitments which have been made.

Aside from the problems with fulfilling committed orders, there is the issue that forecast demand is uncertain. The uncertainty is not evenly distributed, making some of the demand forecast much more questionable than other portions. Because of this, it would not be the best practice to devote all of the available production capacity to the highest profit forecast items. The theory from research for this type of problem suggests a scenario-based, stochastic programming approach. This type of solution strategy is impractical when some of the *deterministic* problem sizes are larger than 100 000 rows. It is also a severe burden to develop scenarios for the user community which is in general unfamiliar with linear programming.

5 THE BPS APPROACH—GOAL PROGRAMMING ON DEMAND CLASSES

A practical and successful approach for even the largest-scale manufacturing problems is to break up overall demand into 'demand classes' reflecting marketing priorities. For example, class 1 might be highest-priority committed orders, class 2 might be regular customer orders, class 3 might be rebuilds of safety stocks, class 4 might be reliable-demand forecasts, and class 5 might be more speculative demand forecasts. Classes 1 and 2 are naturally defined in marketing order entry systems. Class 3 is also naturally defined based on finished goods or intermediate product inventory policies. Classes 4 and 5 at first glance may seem less practical for marketing management to define, but the overall forecast for a product (less committed orders) may be automatically broken into classes 4 and 5 by tracking and analysis of the forecast errors. For example, subtracting one or two standard deviations of forecast error from the forecast could be used to define class 4, and the balance of the forecast could be placed in class 5.

An incremental capacitated-loading calculation is performed, whereby demand classes are loaded in priority order. This also means that requirements planning must be performed for each class. It is computationally more tractable to work with *cumulative* demand classes, so the classes are redefined such that class n includes the demands in classes $1,2,3,\ldots,n$.

Requirements-planning calculations for each (cumulative) demand class are performed. These are relatively small optimization calculations, and the contribution is small to the overall run time from performing multiple requirements-planning calculations, one for each class. In actual implementation, requirements planning for all demand classes of a product family is performed as a single calculation, with the inclusion of consistency constraints between classes, as discussed in Leachman (1993).

Next, capacitated-loading calculations are performed for each class. After solving a particular class (except the last), bounds on backorder variables are introduced to the formulation for the next class to make sure that its solution will not diminish the service level provided by the solution for the previous demand class.

The backorder level of class $n+1$—relative to the demands in class n—is constrained to be less than or equal to the backorder level calculated in class n for each product. This constraint can be enforced in terms of upper bounds on the backorder variables relative to cumulative demands over both classes. The upper bound is then the backorder level from class n plus the difference between the cumulative demands of class $n+1$ minus class n. Bounds are then readjusted after optimality is attained for each class.

Note that the structural formation does not change. Cumulative product demands (cumulative over time) are increased, and feasible bounds on backorder variables are introduced. The optimal solution to class n is a feasible solution to the problem for class $n+1$ if the backorder variables are adjusted upwards by the increment in the demands. The overall approach may be thought of as goal programming, but constraints are not placed on values of multiple objective functions, and no constraints for multiples types or scenarios of demands are carried. Instead, bounds are placed on the variables and the right-hand sides of constraints are modified moving from class to class. This keeps a practical size and a practical run time for the computation.

Differentiation of demand classes into orderboard, inventory, and forecast types is also done to allow different objective functions to be used for each case. All of the factors in the objective functions are discounted to incorporate the 'cost of money' over time. Inventory holding costs are a real-world cost of manufacturing. Their placement in the objective functions keeps the system in line with the principles of JIT manufacturing.

Orderboard and inventory classes use the same objective function. In this case, meeting the commitment is a high priority so the relative profit of the order should not interfere with meeting the demand as on-time as possible. Maximizing revenue is still an objective, but at a slightly lower priority to giving the best service level possible. The objective function used for orderboard and inventory classes is to minimize the cost of backorder plus inventory-holding costs. All cash flows in the objective are negative in this case as the committed order must be filled regardless of the cost of doing so.

In a forecast class of demand, the objective function is changed to maximization of discounted cash flow. Avoidable costs of production are taken into account, and product revenue is directly represented in the function. Inventory-holding costs are also used in this objective function. Therefore the forecast objective function becomes the maximization of revenue minus production cost (as above) minus inventory-holding cost. The production costs used constitute a very conservative approach. Costs are limited to that portion which will only occur if a start is made (e.g. direct materials, subcontracting). This is a small fraction of the total production cost in the semiconductor industry which is a highly capital-intensive enterprise. The reasoning behind this cost selection is that changing the current build schedule will have no effect upon previously purchased equipment. Information relating directly to capacity can be used to help with future capital purchasing decisions.

The bounds created from the solution to the previous demand class are essentially by 'service level', therefore each new demand-class solution can be completely rescheduled using different product routings. (An alternative method might be to use bounds on production variables instead of backorder variables. This would force each succeeding solution to use the same method to meet demand as the previous class. The production plan created from this method will not be able to use available WIP and inventory in the most effective manner.) The forecast objective function can make good use of this flexibility. In tandem with the bounds, this objective helps to realign production among various alternative paths to minimize cycle time, overbuilds, and production costs, and to maximize yield. For those products which might have low yields or poor bin splits, the forecast objective function may not create any new availability. This type of feedback could be extremely useful to both manufacturing and marketing organizations. If the product is not profitable with the very conservative productions costs used here, it will most likely be worse in the true business analysis.

6 IMPLEMENTATION AT HARRIS

At Harris Semiconductor, the planning cycle begins at 00.00 Saturday (Eastern USA time) each week. The time-line goal for completing all of the calculations is 13.00 Sunday. This time includes getting data from all of the production sites. Currently, Harris uses four demand classes which represent: (1) committed orderboard, (2) safety-stock rebuilds, (3) uncommitted orders and quotes, and (4) forecast. At the present time, only the orderboard/inventory objective function is being used. This decision was made to help get over the startup hurdles which are inherent with any new system installation. In this case, the new system is radically different from the previous practices.

This type of change is extremely difficult in a large organization. The use of computer calculated optimization models is an entirely new paradigm to an organization which for the most part relied on manually generated production plans. Linear programming is a method which was practically unknown to Harris planning and manufacturing personnel before the advent of this project. All of these areas have required a large amount of education for both management and the general user community. Company policies with regard to the planning process have had to be revisited and revised and in many cases, new policies have had to be created. As this is written, the changes and education are far from being completed!

7 NUMERICAL SIZE OF HARRIS PROBLEMS AND COMPUTATIONAL EXPERIENCE

Problems were run on an IBM RISC System/6000 Model 560 (256 Mb memory) using OSL to solve the LP problems. Below, elapsed time is expressed as H:MM:SS and CPU time is in seconds. Module 4 problems were run to 'interior optimum' (on a surface, not a vertex) while Module 3 problems use the simplex method and are run to optimality.

7.1 Module 4 (assembly and test)

The problem sizes here are for one of the production facilities, known as KL Plastic, in Kuala Lumpur, Malaysia, see Table 6.1.

7.2 Module 3 (material fab, wafer fab, and wafer probe)

These problem sizes correspond to several US plant sites which build 'IC' wafers see Table 6.2.

Table 6.1 Module 4—interior point method—interior optimum.

Priority class		1	2	3	4
Matrix size	Rows	86212	94638	94721	104312
	Columns	190661	211872	212092	235283
	Entries	680243	744997	745516	814186
Timings	Elapsed	1:06:30	1:58:49	1:59:28	2:24:36
	Total CPU	3933.450	7045.010	7082.640	8560.310
Iterations		50	88	94	110

Table 6.2 Module 3—simplex method.

Priority class		1	2	3	4
Matrix size	Rows	10613	13168	13265	23815
	Columns	30230	37658	37940	69652
	Entries	109388	135155	135895	251410
Timings	Elapsed	0:03:21	0:02:28	0:00:57	0:27:49
	Total CPU	197.4000	143.7900	53.58000	1651.890
Iterations		7300	3625	286	17643
Advanced basis		N/A	Yes	Yes	No

8 RESULTS

Subsequent to the implementation of IMPReSS, on-time delivery performance at Harris's semiconductor sector climbed steadily from 75% before implementation to a plateau of 95% during all of 1993. The semiconductor sector has moved from losses to steadily increasing profits. This performance is consistent with the design principles of BPS: production plans are prepared so as to ensure on-time performance against previous-customer commits as a first priority, then filling out the plan so as to load the availability with the maximum discounted cash flow considering the available capacity and sales forecasts. While the IMPReSS system is by no means the only improvement to semiconductor operations implemented over the past several years at Harris, it clearly is a major contributor. Encouraging as these results may be, in our opinion, IMPReSS still has some ways to go to achieve full acceptance at Harris and allow the company to fully reap the potential benefits.

REFERENCES

Leachman, R. C. (1993) Modeling techniques for automated production planning in the semiconductor industry, in: T. A. Ciriani and R. C. Leachman (Eds), *Optimization in Industry—Mathematical Programming Methods and Modeling Techniques in Practice*, John Wiley and Sons, Chichester, pp. 1–30.

Robert C. Leachman
Engineering Systems Research Center
University of California
Berkeley
CA 94720
USA

Dale J. Raar
Harris Corporation—Semiconductor Sector
Manufacturing Systems, MS 62A-024
2401 Palm Bay Road
Palm Bay
FL 32905
USA

7
Optimizing the Siting of Depots and their Input–Output Capacities

Paolo Migliavacca
Enichem Agricoltura, Milan, Italy

ABSTRACT
This study sets out to cast light, by means of a tensorial system, on the optimal siting of depots according to: their maximum stock capacities; their packaging capacities (sacks, pallets, bales, etc.); input–output; the optimal number of stock sites. Following this optimization, the study proceeds to the calculation of the increased costs associated with the use of only already existing depots.

1 INTRODUCTION

The concept of logistical and stock optimization is important because of the operating needs due to the following factors:

- Customers require prompt supply.
- The factory warehouses cannot meet peak demands.
- The product deteriorates if stored for long periods.
- Transport costs are less if the depot stocks are replenished evenly throughout the year.

The last ten years have therefore seen the development of a series of optimizations, at first partial and then ever more complex, covering manufacturing activity, logistics, delivery times, and stock levels.

The peninsula of Italy, some 900 km (550 miles) long and between 100 and 225 km (60 and 140 miles) wide, is split by the Apennines, a range of mountains

Optimization in Industry 2. Edited by T. A. Ciriani and R. C. Leachman

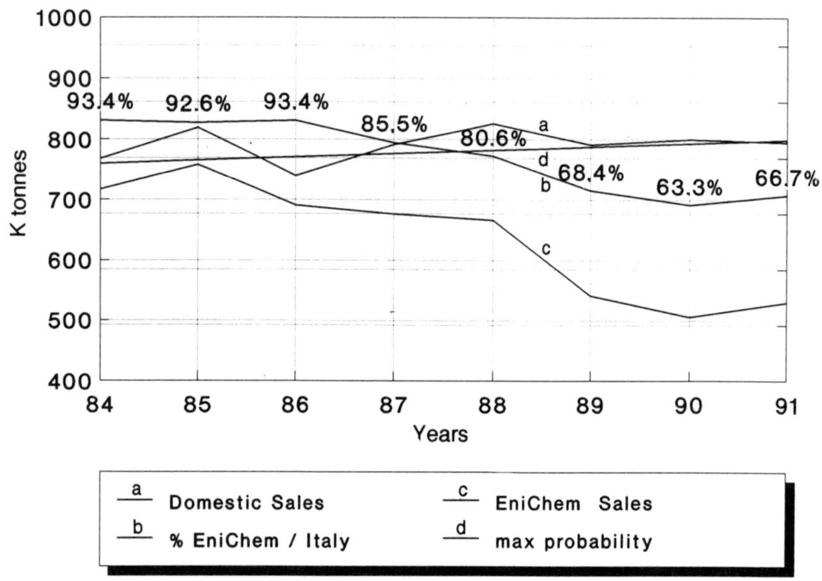

Figure 7.1 Calcium ammonium nitrate: Italian market.

rising to almost 3000 m (9500 feet). For administrative purposes, it is divided into 19 regions (including the major islands of Sicily and Sardinia) and 85 provinces.

2 HYPOTHESES AND INPUT DATA

In order to meet the demand for fertilizers, which is concentrated in ever shorter periods, it became clear that it was essential to develop an organic and coordinated policy and procedure that was able, at any moment, to keep abreast of market fluctuations and to exploit any openings for growth.

2.1 The Italian market

The market was considered in relation to sales by product, by province, and by month. In order to pinpoint further sales possibilities, the historical data on consumption from 1984 was analyzed. Harmonic disturbances were filtered (Epstein and Zin, 1989) out of this data, provided by the Italian statistical office ISTAT. See Figures 7.1 to 7.6. The data for typical annual sales, furnished by the sales function in Italy, was combined with the official statistics to permit a study of the potential market, its evolution, and possible medium-term (3–5 years) variations. This study has permitted the identification of possible marginal variations caused by increasing production of certain crops and/or changes in traditional consumption which could result in a more aggressively competitive market.

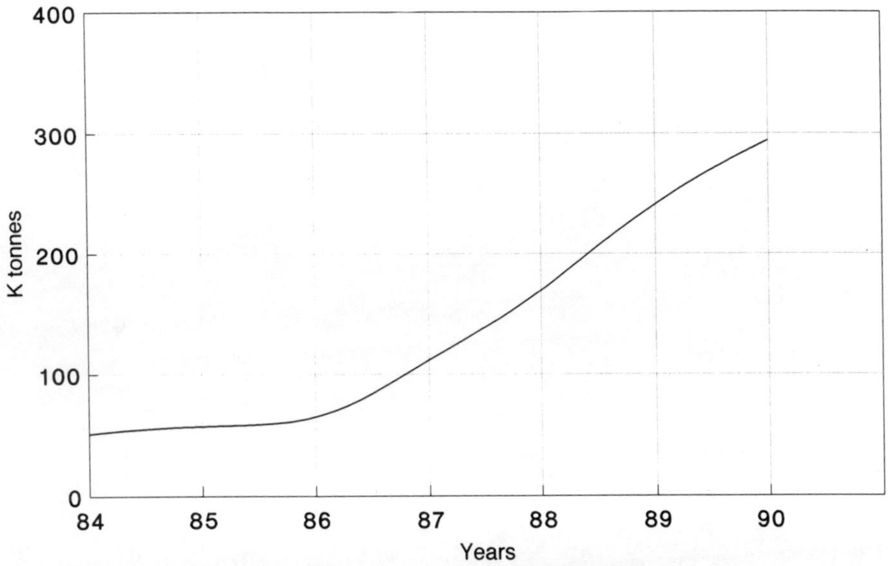

Figure 7.2 Calcium ammonium nitrate: market differential.

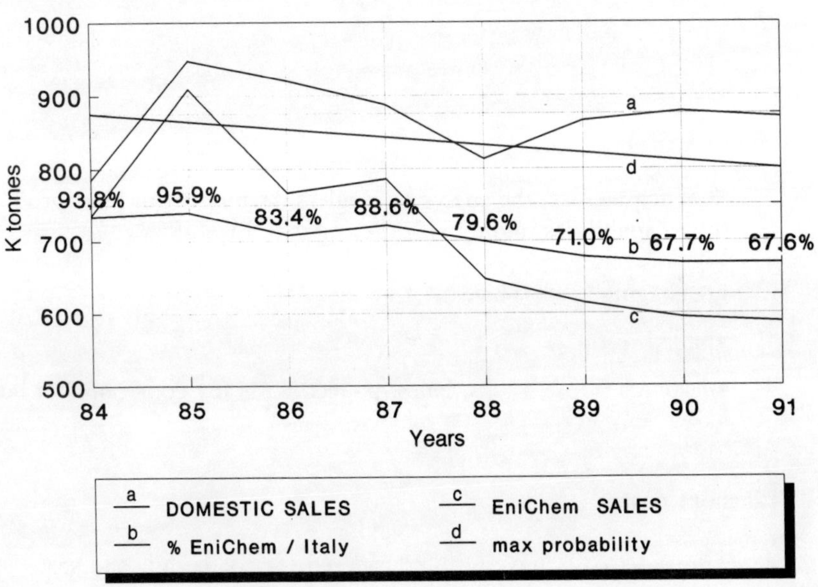

Figure 7.3 Urea: Italian market.

2.2 The foreign market

The budget forecasts foreign sales by month, product, market, and country; such sales are marginal to the home market. The European market is comparable as a marginal contribution by product to the Italian market.

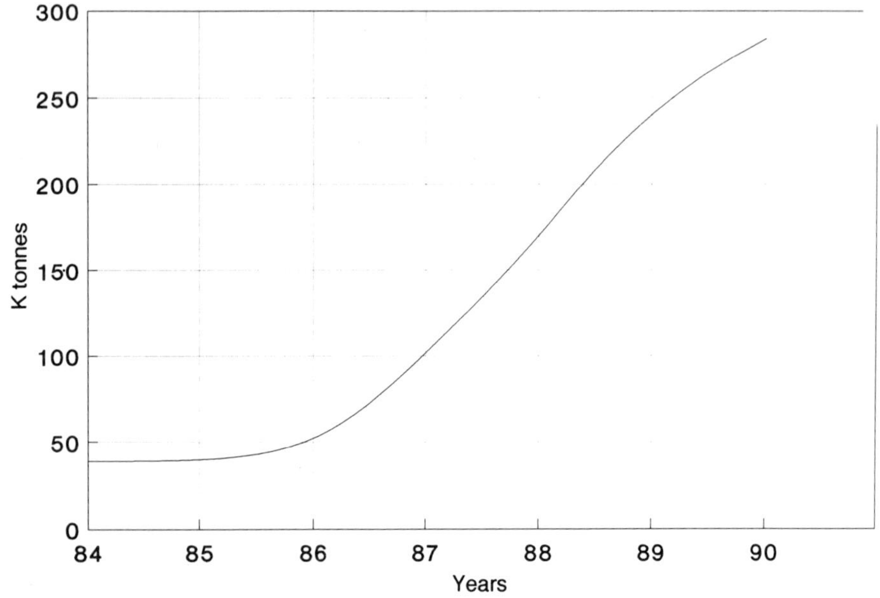

Figure 7.4 Urea: market differential.

2.3 Factory depots

For each factory, the analysis considers the number of bins (product in bulk) and their capabilities, along with the capacity by shift for:

- packaging (sacks, bales),
- palletization,
- despatch (loose, sacks, bales, pallets) with any constraints of interdependence.

2.4 Transport costs

Transport costs have been considered for road, rail, and sea.

Road transport

A road network model has been developed with the distances in kilometers between the factories, from each factory to each province throughout Italy, and from each province to every other province. The latest calculations use tariffs per kilometer as of 1 June 1991.

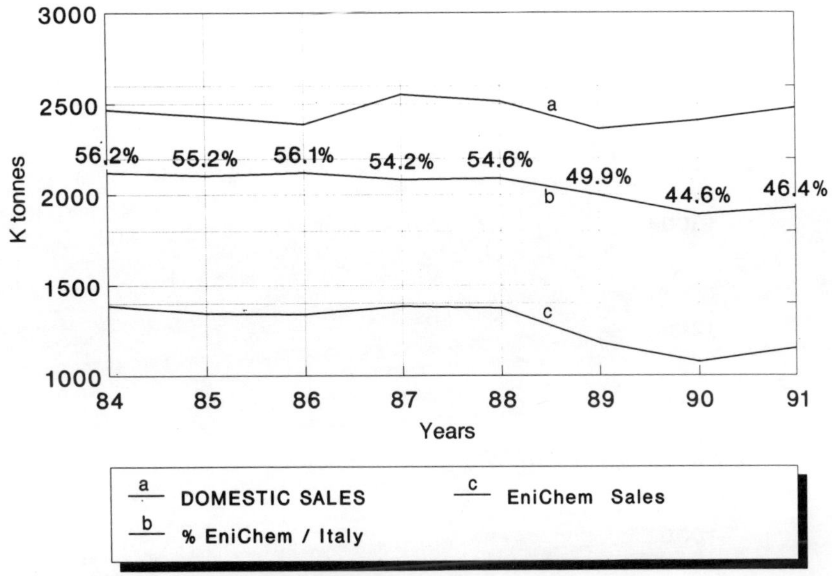

Figure 7.5 Compounds: Italian market.

Rail transport

A similar method has been used for rail transport, taking into account the preferred routes. Rail transport is not considered for journeys of less than 100 km (60 miles). At this stage, rail transport is considered only for supplying the depots from the factories; consignments from depots to retail outlets may be considered later.

Sea transport

A network of possible harbors on the entire coastline of the peninsula has been set up, with details of unloading facilities and logistical data.

3 THE TENSORIAL MODEL

3.1 Initial stock

The purpose of the study is to determine the optimal stock levels at January. A model of 12 periods (months) simulates monthly sales and production on the hypothesis that the factories have infinite capacity for stock and despatch (that is, no constraints of maximum stock, number of warehouse bins, despatch

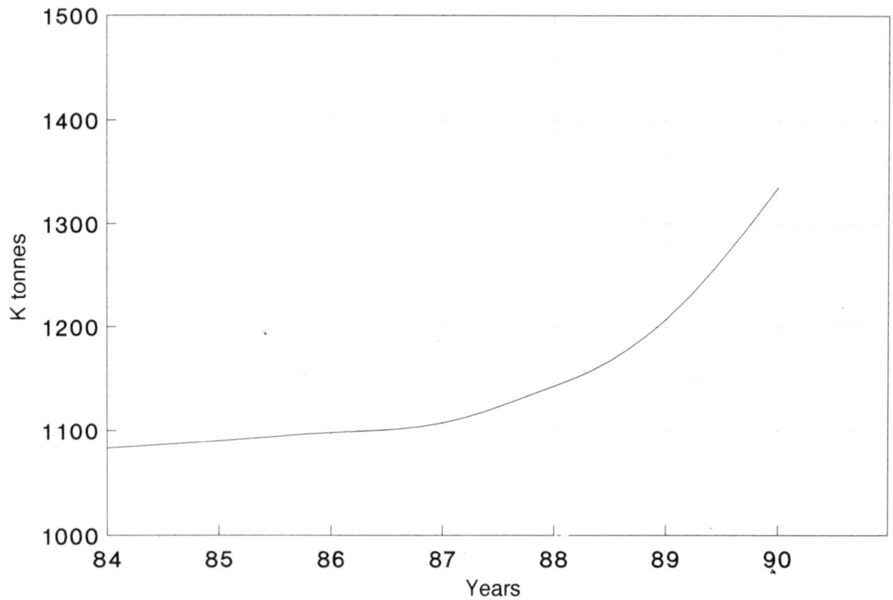

Figure 7.6 Compounds: market differential.

capacity). The model was 'closed' on itself and the optimization stopped when the change in stocks between two solutions dropped below 2%; optimization also stopped after 83 solutions (years). The stock volumes thus calculated (Figure 7.7) were used as initial and final levels in a follow-on algorithm to define the depots.

3.2 The tensorial system

The sales by product, method of transport, and region became the parameters of consumption by product, month, province, and region. The result is a quantity by product, month, and province; this quantity is simulated as the intensity of a force applied at the province.

With the introduction into the model of factory constraints (capacity for stocks and despatch by method of transport), the optimization makes it possible, on the basis of sales and of the costs of stocking input–output, replenishment, and transport to the destination, to determine a series of centers of gravity (depots) so as to minimize costs, both variable and fixed, these last being treated as Boolean variables in the optimization.

The logical consequences of this procedure are:

- medium–high use throughout the year of warehouse space;
- the use of bulk trains or, where possible, ships for replenishment of depots;
- the rational use of the various methods of transport of despatch and means of transport;

PRODUCTION	I	II	III	IV	V	VI	VII	VIII	IX	X	XI	XII	Σ
P.MARGH.	41.4	38.5	42.5	39.7	42.2	38.5	38.3	22.1	39.1	42.5	39.9	42.8	467.5
RAVENNA	45.3	40.6	45.0	43.9	45.3	29.0	33.0	35.0	33.7	44.5	43.9	45.3	484.5
Σ	86.7	79.1	87.5	83.6	87.5	67.5	71.3	57.1	72.8	87.0	83.8	88.1	952.0

- SALES	I	II	III	IV	V	VI	VII	VIII	IX	X	XI	XII	Σ
DOMESTIC	71.0	101.8	106.0	58.3	53.5	46.1	20.6	9.6	10.6	9.6	10.3	32.3	530.0
EXPORT	20.0	20.0	20.0	40.0	40.0	40.0	10.0	22.0	50.0	60.0	60.0	40.0	422.0
Σ	91.0	121.8	126.0	98.3	93.5	86.1	30.6	31.6	60.6	69.6	70.3	72.3	952.0

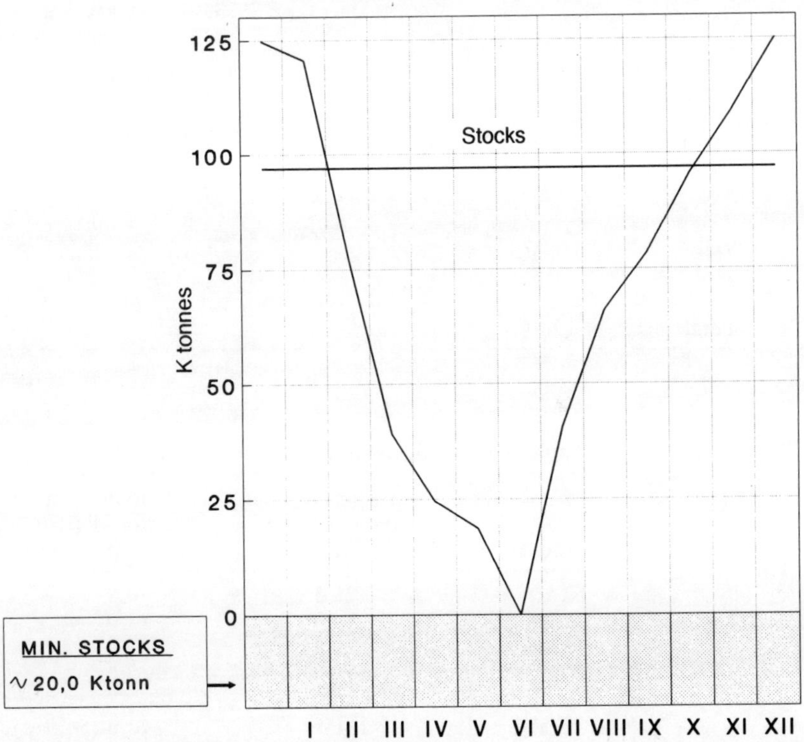

Figure 7.7 Calcium ammonium nitrate 26/27.

• the method and means of transport for minimum costs and maximum profits in the areas of production, distribution, and sales.

Appendix A formally presents the optimization model.

4 THE OPTIMIZATION

The solution has shown clearly that the most economical arrangement is five mainland depots, plus the two depots in Sicily and Sardinia. The optimal pattern of replenishment has been calculated as in Table 7.1.

Table 7.1 Optimal replenishment pattern.

Depot	Kt/year	Peak month (deliveries)	Product	Ktonnes	Sacks	Bales	Pallets	Loose
Piacenza	200	April	Super	15.0	—	1.5	13.5	—
			CX	30.0	—	3.0	27.0	—
			Total	45.0	—	4.5	40.5	—
Massa	150	October	CX	10.0	5.0	—	5.0	—
			18/46	20.0	5.0	5.0	10.0	—
			Total	30.0	10.0	5.0	15.0	—
S. Benedetto del Tronto	200	November	0/46	5.0	1.0	1.0	3.0	—
			18/46	30.0	3.0	3.0	24.0	—
			Total	35.0	4.0	4.0	27.0	—
Salerno	150	September	N.A.	3.0	2.0	—	1.0	—
		April	Super	8.0	6.4	—	1.0	0.6
			CX	11.0	7.0	—	3.0	1.0
			Total	22.0	15.4	—	5.0	1.6
Barletta	200	September	N.A.	5.0	3.0	—	1.0	1.0
			Super	15.0	5.0	—	2.0	8.0
			CX	35.0	10.0	—	3.0	2.0
			Total	55.0	18.0	—	6.0	11.0
Sicily	150							
Sardinia	100							
Total	1150							

Table 7.2 Products in the Barletta depot.

Product	Total consumption in the catchment area (Ktonnes)	Amount supplied from Barletta (Ktonnes)
N.A.	91.5	50.0
Super	50.0	25.0
18/46	75.0	45.0
CX	45.0	20.0
Total	261.5	140.0 (54.0%)

Table 7.3 Breakdown of supplies (see Figures, 7.8, 7.9, 7.10).

	N.A.	Super	CX	18/46	Total
January	20.0	5.0	—	5.3	25.0
February	15.0	5.0	—	5.3	25.3
March	15.0	5.0	—	10.7	30.7
April	—	5.0	—	20.3	25.3
May	—	—	10.0	8.6	18.6
June	—	5.0	10.0	0.1	15.1
Total	50.0	25.0	20.0	45.0	140.0

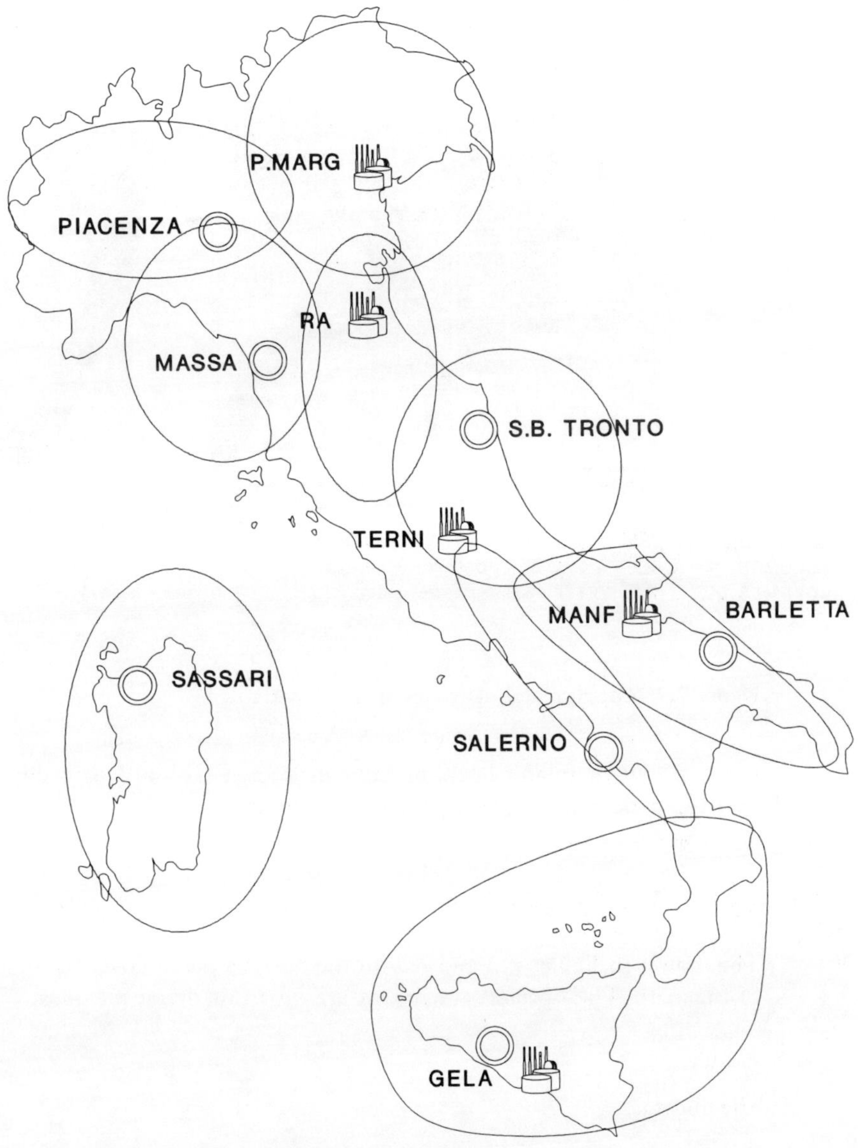

Figure 7.8 Optimal catchment for distribution depots.

The use of depots increases distribution costs by about 9300 million lire (about $5.8 million) as compared with the hypothetical costs of distribution direct from the factories (which is not, however, possible), but has the following advantages:

- the possibility of coping with concentrations of demand;
- a tolerance of about two weeks at times of greatest distribution;
- the ability to meet unexpected demand anywhere in Italy at short notice;

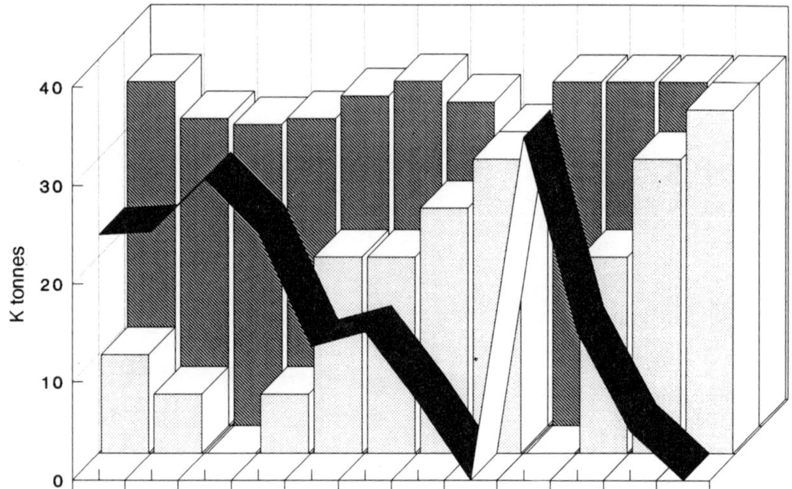

	1	2	3	4	5	6	7	8	9	10	11	12
TOTAL TONNAGE	35	31	31	31	34	35	33	30	35	35	35	35
VOLUMES IN	10	6	0	6	20	20	25	30	0	20	30	35
VOLUMES OUT	25	25	31	25	14	15	8	0	35	15	5	0

Months

Figure 7.9 Supplies and deliveries at the Barletta depot.

- throughout mainland Italy, orders can be met within 2 or 3 days, as against 10 to 15 days from the manufacturing plants.

4.1 An example: The Barletta depot

The figures in Tables 7.2 and 7.3 for the Barletta depot cover the first six months of the year. The optimal catchment area for this depot includes:

- Pulia,
- Basilicata,
- the northern part of Calabria,
- part of Abruzzo.

The products handled by Barletta are given in Table 7.2.

4.2 Cost of using non-optimal depots

The simulation allows us to force another site to replace one of the optimal sites. Keeping the optimal solution for purposes of comparison, the effects of alternative sites were analyzed.

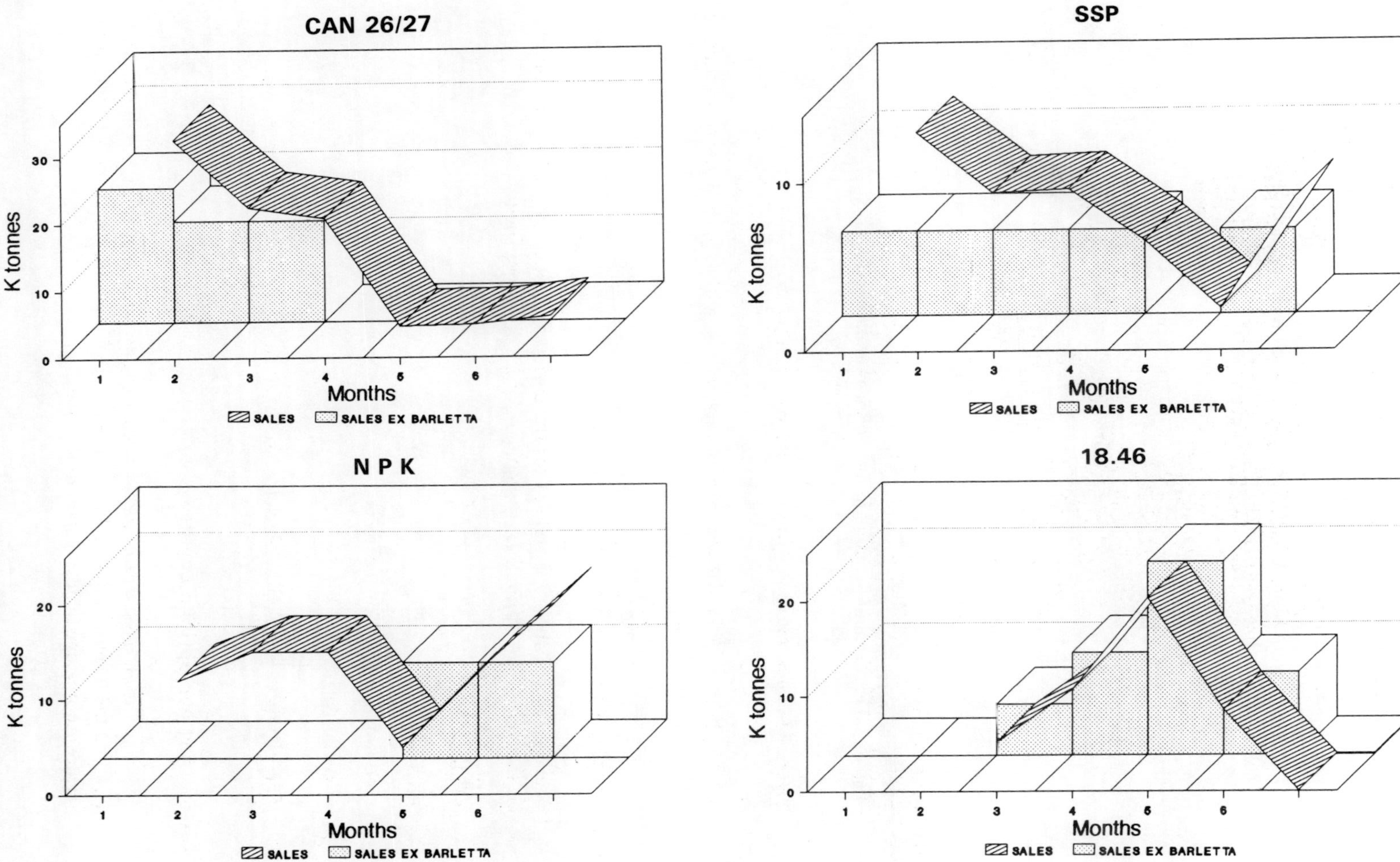

Figure 7.10 Sales by product in the Barletta catchment.

- If Piacenza were replaced by Cerea (Verona) and Agrilombardia, the increased transport costs would be 9000 million lire (about \$5.6 million)—4.5 lire per kilo (\$2.8 per metric ton).
- If Barletta were replaced by Agrilevante, the increased transport costs would be 2600 million lire (\$1.6 million).

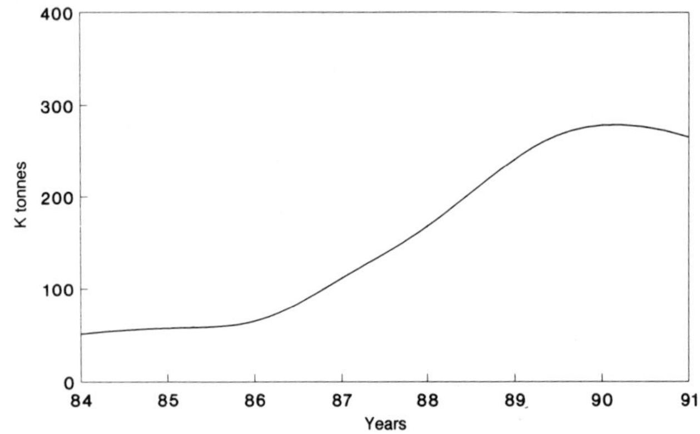

Figure 7.11 Calcium ammonium nitrate: market differential.

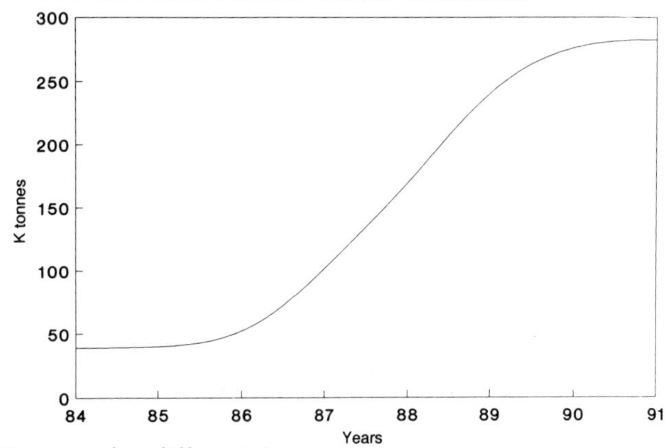

Figure 7.12 Urea: market differential.

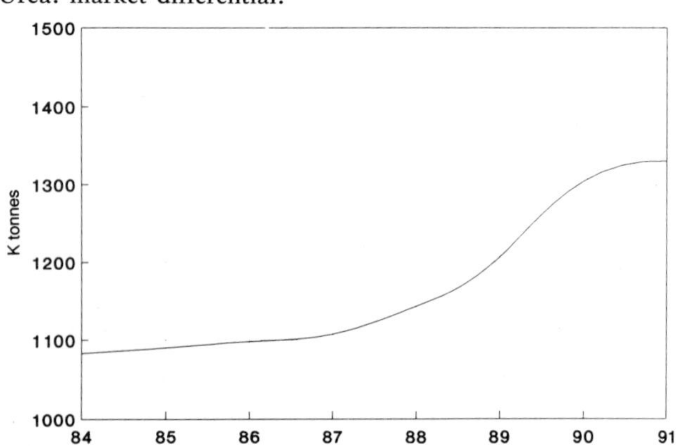

Figure 7.13 Compounds: market differential.

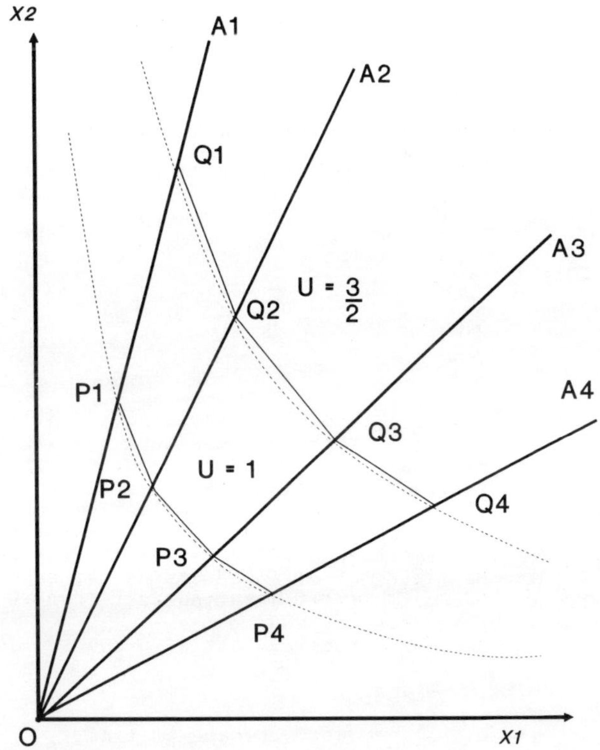

Figure 7.14 Indifference map for two arbitrary depots.

Appendix B explains the methodology for computing the increased costs of using non-optimal depots

5 POSTSCRIPT

This experimental study began in 1991 and was completed one year later when the final data relevant to that year had become available. The study gave evidence (Figures 7.11, 7.12, 7.13) of two positive results from a more aggressive marketing policy on the one hand and a smoother and quicker delivery pattern due to appropriate location of the warehouses on the other:

- The slowing down of imports allowed a recovery of marginal market shares.
- The invoiced value showed an average increase during the entire period considered. This was obtained with a pricing policy that did not take into account pre-storage prices since the above-mentioned location of warehouses had a buffering effect on peak-demand periods.

APPENDIX A

EasyModeler (IBM, 1993) notation is used to formulate the mixed integer model.

```
Depo_Site GENERIC          |*  Depo-Site Optimization Model  *|

SET
 i Source;
 j Destination;
 t Time;
 p Product;
 m Transportation_media;
ENDSET

DATA
 A[i,t];           |*  Source i production investment availability at time t  *|
 B[i,t];           |*  Source i stocking investment availability at time t    *|
 D[j,t,p];         |*  Destination j demand of product p at time t            *|
 Q[i,t,p];         |*  Source i avalailability of product p at time t         *|
 G[i,j,t,p,m];     |*  Transportation media m selection weight at time t      *|
                   |*    to transport the product p from i to j               *|
 H[i,j,t,m];       |*  Transportation media m availability at t from i to j   *|
 U[i,t];           |*  Max level of all products delivered from i at time t   *|
 C[i,j,t,p,m];     |*  Product p transportation cost from i to j by m at t    *|
 F[i];             |*  Fixed cost of source i                                 *|
 R[i,t,p];         |*  Source i stocking cost of product p at time t          *|
ENDDATA

INTERNALDATA
                   |*  Global product p demand at time t                      *|
 E[t,p]= SUM(D[j,t,p])(OVERALL j);
ENDINTERNALDATA

VARIABLE
 x[i,j,t,p,m];     |*  Product p flow from i to j at time t by media m        *|
 w[i,t,p];         |*  Delivered product p at time t from source i            *|
 s[i,t,p];         |*  Stocked product p at time t in source i                *|
 z[i,t];           |*  Total flow from source i at time t                     *|
 y[i] LOGIC;       |*  =1 if source i is active; =0 otherwise                 *|
ENDVARIABLE

CONSTRAINT
 Product_Flow (FORALL i,t,p) SUM (x[i,j,t,p,m]) (OVERALL j,m) = w[i,t,p];
 Product_Investment (FORALL i,t) SUM(w[i,t,p])(OVERALL p) <= A[i,t]*z[i,t];
 Stocking_Investment (FORALL i,t) SUM(s[i,t,p])(OVERALL p) <= B[i,t]*z[i,t];
 Starting_Status (FORALL p, FOR t=0) SUM(w[i,t,p])(OVERALL i)
    =E[t,p]+SUM(s[i,t,p])(OVERALL i);
 Continuity (FORALL p, FOR t>0) SUM(w[i,t,p])(OVERALL i)
    +SUM(s[i,t-1,p])(OVERALL i) = E[t,p]+SUM(s[i,t,p])(OVERALL i);
 Product_Availability (FORALL i,t,p) w[i,t,p]) <= Q[i,t,p]*y[i];
 Media_Selection (FORALL i,j,t,m) SUM(G[i,j,t,p,m]*x[i,j,t,p,m])(OVERALL p)
    =< H[i,j,t,m];
ENDCONSTRAINT

LIMIT
 Max_Investment (FORALL i,t) z[i,t] <= U[i,t];
ENDLIMIT

OBJFUNCTION
                   |*  Transportation, depo set-up and stocking expenses      *|
 MINIMIZE Cost SUM(C[i,j,t,p,m]*x[i,j,t,p,m])(OVERALL i,j,t,p,m)+
     SUM(F[i]*y[i])(OVERALL i)+ SUM(R[i,t,p]*s[i,t,p])(OVERALL i,t,p);
ENDOBJFUNCTION
```

APPENDIX B INCREASED COSTS FOR THE USE OF ALREADY EXISTING DEPOTS

The increased costs are considered as the change in the indifference map of the two depots. For simplicity, the interdependence between two depots (X_1 and X_2) can be illustrated two-dimensionally showing the fixed investments of the two depots with a matrix whose unitary level of each activity is defined to give utility $u = 1$ (indifference).

Coefficients of investments:

$$A = \begin{vmatrix} a_{11} & a_{12} & a_{13} & a_{14} \\ \\ a_{21} & a_{22} & a_{23} & a_{24} \end{vmatrix}.$$

Level of utility (u): 1 1 1 1.

The level of indifference is formed by four points P_r with coordinates (a_{1r},a_{2r}) for $r = 1,2,3,4$ together with their combinations. Figure 7.14 shows the shape of the level of indifference.

The increase in cost for the substitution of the two depots is given by λ of order $(n + h)$. The linear program sets out to minimize the costs, and so we have a negative λ. The dual program seeks for a positive λ but, because the second depot is a constraint, emphasizes the negative value of the new solution.

REFERENCES

Epstein, L., and Zin, S. (1989) Substitution, risk, aversion and temporal behavior, *Econometrica*, 57, 937–969.

IBM (1993) *AIX EasyModeler/6000, User Guide* SB13-5249, IBM Corporation.

Enichem Agricoltura
Via Medici del Vascello 40/C
20100 Milan
Italy

8
Demand Allocation for Capacity Optimization: A Solution in IBM Manufacturing

Christof Dillenberger and Artur Wollensak
IBM German Manufacturing Technology Center (GMTC), Sindelfingen, Germany

ABSTRACT
Reducing costs and avoiding investments while increasing the productivity—these are today's needs for manufacturing industry. This goal can only be achieved with great effort and the use of advanced methods. Mathematical programming is one of these methods. The first part of the chapter presents a decision-support system which is used successfully in IBM manufacturing. The second part is devoted to the discussion of different 0–1 models for period-overlapping setups in multi-period production planning.

1 INTRODUCTION

Loss of productivity in manufacturing has many reasons such as product-mix fluctuations, unexpected machine breakdowns, varying processing times, machine setups, unfavorable machine assignments, and poor production sequences. It is necessary to consider all these factors when attempting to increase productivity. Human experts are not able to do this, but mathematical approaches can handle these problems. Today, there exist many linear and mixed-integer programming models to solve various complex production-planning problems very efficiently (Tempelmeier and Kuhn, 1992). But only some of the methods are actually used in practice because not all are incorporated in appropriate software tools (Fleischmann, 1988).

At the IBM German Manufacturing Technology Center (GMTC) we developed and implemented the decision-support system DACAPO (Demand Allocation for Capacity Optimization). It is a user-friendly tool and is successfully used by manufacturing people in various applications.

Optimization in Industry 2. Edited by T. A. Ciriani and R. C. Leachman

The goal of this chapter is twofold:

1. To show the working of DACAPO on a specific application. This application deals with the optimal assignment of products on different machines within a micro-mechanics manufacturing line.
2. To discuss four different 0–1 model approaches to consider setups in multi-period production planning. Various production-planning problems can be described and solved with linear programming (LP) models. In manufacturing applications machine setup often plays an important role. To solve a setup problem 0–1 variables must be introduced in the model, which leads to an increase in the model size and computational effort.

2 AN APPLICATION

At the IBM plant at Sindelfingen complex components for mainframe computers are manufactured in a microelectronics production line (Fromm *et al.*, 1993). In a specific process sector material is carried off by a laser-ablation process. Five laser-ablation tools (hereafter 'lasers') are used to produce four different products in a three-shift environment. The shift manager's continuous task is to decide: which product should be assigned to which laser to meet the given demand for the shift or day. This task is not easy because of the varying processing times of the different products on the different lasers (see Table 8.1).

This heterogeneous processing-time matrix is caused by the different technical capabilities of the lasers and the different processing requirements of the product. An additional effort of 30 minutes is necessary to set up a laser if the product changes.

Let us assume that the five machines are set up for the products as in Table 8.2. From the logistics system the shift manager gets the information about

Table 8.1 Processing time per part (in minutes).

	BSM	FDM	LVL2	LVL5
Laser1	15.0	20.0	15.0	—
Laser2	—	—	15.0	17.0
Laser3	10.0	18.0	15.0	—
Laser4	13.0	18.0	15.0	15.0
Laser5	20.0	17.0	15.0	21.0

Table 8.2 Setup per machine.

Laser1	FDM
Laser2	LVL2
Laser3	LVL2
Laser4	LVL2
Laser5	LVL5

Table 8.3 Demand per product.

BSM	186
FDM	18
LVL2	167
LVL5	62

the production volume of the day (see Table 8.3): because of an urgent customer demand it is necessary to produce all 18 parts of the product FDM.

The shift manager tries to reduce the setup times when assigning the products to the machines. This should result in a high machine utilization. The assignment rule is as follows. Assign to a specific machine the product for which the machine is currently set up. If the demand for this product is satisfied, then change the product on the machine. Choose the product with the shortest processing time on this machine. When assigning the given demand according to this rule, the production sequence shown in Figure 8.1 results. This figure shows that five setups are necessary and from the total demand of 433 parts a volume of 52 (or 12%) could not be produced within the three shifts of the day.

This leads to the question as to whether it is possible to satisfy the entire demand with a different assignment. Using a system like DACAPO (where the goal is maximize throughput and minimize setup effort) the optimal assignment as shown in Figure 8.2 is generated. Please notice that, even with two additional setups, it is possible to fulfill the entire demand within three shifts.

This relatively small sample shows that an unfavorable assignment may lead to large amounts of wasted machine capacity. In practice often additional constraints (e.g. minimum production volumes for specific products) must be considered. Here the human expert is often not able to solve these problems. Systems like DACAPO can support human decisions and enable the people to react very fast to changing production situations. The decision-support system DACAPO is for example used to help to answer the following questions:

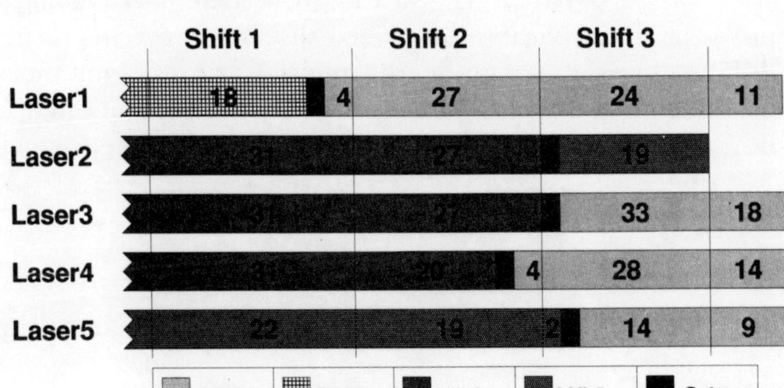

Figure 8.1 Manual Assignment.

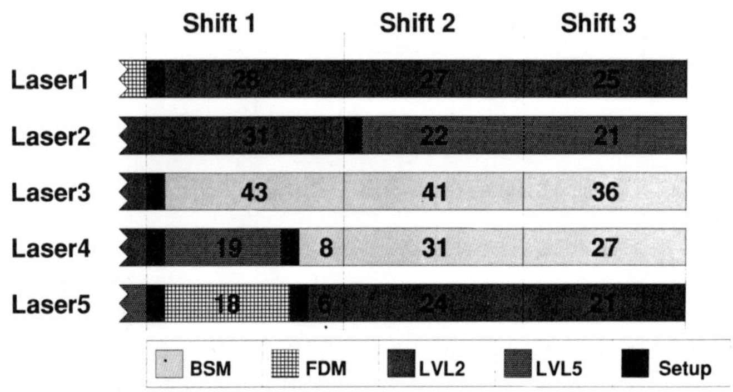

Figure 8.2 Optimal Assignment.

- A machine breakdown occurs and lasts for a long period. What are the consequences for the throughput? Is it necessary to change the machine assignments?
- Customer demand reduces temporarily. Is it possible to shut down a machine to perform a preventive maintenance?
- Customer demand increases. Are the capacity reserves sufficient?
- The product mix changes. Are the capacity calculations still correct? When is it necessary to increase capacity?

The 10% capacity increase in the above example was also achieved in other applications (Wollensak *et al.*, 1992). Also, we achieved

- shorter cycle times,
- early detection of low- and high-load situations, and
- reduced work-in-process.

Depending on the machine costs this results in savings of up to 1 million DM per year. The use of DACAPO also leads to a better understanding of the manufacturing process. The application described shows one specific scenario. Others include drilling of printed circuit boards, punching of foils, and the operation of flexible manufacturing systems. DACAPO is very flexible and can be used very efficiently in production centers which are characterized in the following way:

- a high part-time mix,
- a jobshop-like machining environment,
- parallel machines with different technical capabilities (e.g. functions, speed, availability),
- flexibility in machine assignments, or
- part-type-dependent setup times.

We now come to the second part of the chapter which discusses different 0–1 models for period-overlapping setups.

3 SETUP APPROACHES

There are many different models to attack production-planning problems. Most of them are formulated as linear programming (LP) models. But in manufacturing period-overlapping setups often plays an important role.

The modeling of setup requirements leads to the introduction of 0–1 variables since we have to find out whether a setup is necessary or not. In fact this is a typical situation for the use of 0–1 variables. We compared four different cases with respect to the modeling accuracy and resulting problem size:

1. One approach is to split the user-defined periods into so-called micro-periods of equal length (see Figure 8.3). For each micro-period we allow only one single part type (A, B, or C). From the modeling point of view this approach is very easy to handle, but it has some drawbacks. The major one is the estimation of the optimal number of micro-periods. If this number is too large, the size of the 0–1 model will explode, since the number of 0–1 variables is proportional to the number of micro-periods (and machines and part types). If the number is too small, we may waste capacity, since the production volumes do not fit exactly into the micro-periods. The waste of capacity is especially high for cases with short processing times or small production quantities.
2. The drawback of wasted capacity in approach 1 can be lessened by allowing variable-length micro-periods (see Figure 8.4). But even in this case the determination of the number of micro-periods is still unresolved.
3. An alternative is to stay with the user-defined periods, but allow multiple part types. A simple approach is always to define a setup for each part type per period (see Figure 8.5). From the modeling point of view this is very easy to handle, but it creates setups where in practice no setup is required (e.g. part types B and C in Figure 8.5). This results in wasted capacity, too.

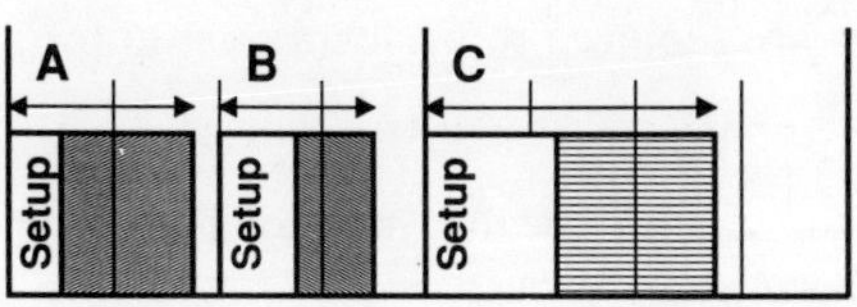

Figure 8.3 Case 1. Two user periods split into 4 fixed length micro-periods. Only 1 part type (setup and/or processing) per micro-period is allowed.

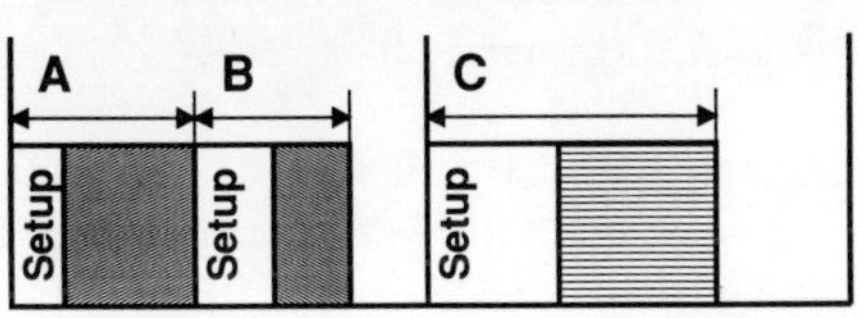

Figure 8.4 Case 2. Two user periods with up to 3 variable length micro-periods. Only 1 part type (setup and/or processing) per micro-period is allowed.

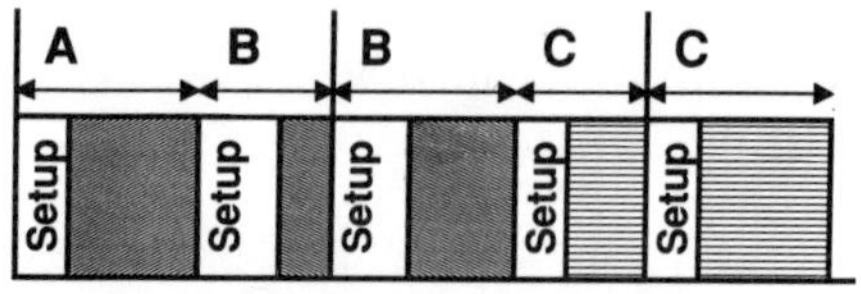

Figure 8.5 Case 3. User periods, multiple part types allowed per period and always setup for each part type assigned to a period.

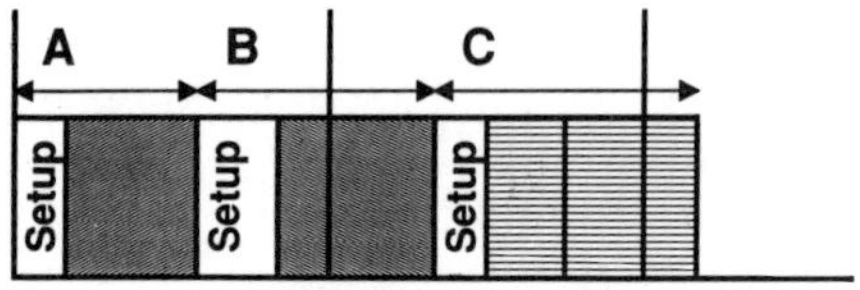

Figure 8.6 Case 4. User periods, multiple part types allowed per period and setup only when necessary.

4. Our best approach allows a part type to be assigned to the same machine in two consecutive periods without requiring machine setup for the second period. In Figure 8.6 the same production quantities are scheduled as in Figure 8.5, but only one setup is considered for part types B and C.

The approach number 4 requires more 0–1 variables than any other, but it is very close to reality and the resulting model size is still unacceptable.

Before we present the model for this approach we introduce the following notation. (Here, we present only a simplified model—the complete model is described in Dillenberger *et al.*, 1994)

T, set of time periods $\{t\}$ in the planning horizon to be studied.

I, set of part types $\{i\}$.

M, set of machine groups (MG) $\{m\}$.

D_i, demand of part type i to be satisfied along the planning horizon.

$p_{i,m}$, processing time required for producing one unit of part type i in MG m.

$C_{m,t}$, capacity of machine m in period t.

$s_{i,m}$, number of time units that are required to perform a setup for any part type i at any machine of MG m.

$\delta_{i,m,t}$, 0–1 variable such that the value 1 means that part type i can be assigned to machine m in period t; otherwise, it is zero.

$\psi_{i,m,t}$, 0–1 variable such that the value 1 means that a setup is required for part type i at machine m in period t and, otherwise, it is zero.

$u_{i,m,t}$, 0–1 variable such that the value 1 means that part type i will be the latest part type in the processing sequence at machine m in period t and, otherwise, it is zero.

$z_{m,t}$, 0–1 variable such that the value 1 means that two part types at least are assigned to machine m at period t and, otherwise, it is zero.

The model must provide a value for the following decision variables:

$x_{i,m,t}$, production volume of part type i that is assigned to machine m in period t.

y_i, backlog of part type i.

Now we are able to formulate the mixed-integer programming model. The target is to obtain the values for $x_{i,m,t}$ $\forall i \in I,\ m \in M,\ t \in T$ and y_i $\forall i \in I$, such that the backlog and processing time along the planning horizon is minimized:

$$\min \sum_{i \in I} y_i + \sum_{i \in I} \sum_{m \in M} \sum_{t \in T} p_{i,m} x_{i,m,t} \tag{1}$$

subject to:

Force $\delta_{i,m,t}$ to have the value 1 whenever part type i is assigned to machine m and period t (constraint (2) is the most frequent usage of 0–1 variables in production planning):

$$x_{i,m,t} \leqslant \overline{M}\delta_{i,m,t} \qquad \forall i \in I,\ m \in M,\ t \in T. \tag{2}$$

Fulfill the demand per product:

$$\sum_{m \in M} \sum_{t \in T} x_{i,m,t} + y_i = D_i \qquad \forall i \in I. \tag{3}$$

Prevent violation of the reduced machine capacity because of setup time:

$$\sum_{i \in I} (s_{i,m} \psi_{i,m,t} + p_{i,m} x_{i,m,t}) \leqslant C_{m,t} \qquad \forall m \in M,\ t \in T. \tag{4}$$

Identify the part type (if any) that causes a sequence-overlapping setup at a machine in a period:

$$\delta_{i,m,t} \leqslant \psi_{i,m,t} + u_{i,m,t-1} \qquad \forall i \in I,\ m \in M,\ t \in T. \tag{5}$$

Define the latest part type in the processing sequence at a machine in a given period:

$$\sum_{i \in I} u_{i,m,t} = 1 \qquad \forall m \in M,\ t \in T, \tag{6}$$

$$u_{i,m,t} \leqslant \delta_{i,m,t} \qquad \forall i \in I,\ m \in M,\ t \in T. \tag{7}$$

Define the number of part types that are assigned to a machine in a given period, such that variable $z_{m,t}$ will take the value 1 whenever more than one part type is assigned to machine m in period t:

$$\sum_{i \in I} \delta_{i,m,t} \leqslant 1 + \overline{M} z_{m,t} \qquad \forall m \in M, \ t \in T, \tag{8}$$

where $\overline{M}$ is a big number, e.g. $\overline{M} = |I|$.

Prevent a part type being assigned as the 'last' part type to be processed at a machine in a given period, if the same part type was assigned the 'last' part at the previous period and at least some other part type has also been assigned to the machine in the current period:

$$u_{i,m,t} + u_{i,m,t-1} \leqslant 2 - z_{m,t} \qquad \forall i \in I, \ m \in M, \ t \in T, \tag{9}$$

$$x_{i,m,t} \geqslant 0, \ y_i \geqslant 0 \qquad \forall i \in I, \ m \in M, \ t \in T, \tag{10}$$

$$\delta_{i,m,t}, \ u_{i,m,t}, \ z_{m,t} \in (0,1) \qquad \forall i \in I, \ m \in M, \ t \in T, \tag{11}$$

$$0 \leqslant \psi_{i,m,t} \leqslant 1 \qquad \forall i \in I, \ m \in M, \ t \in T. \tag{12}$$

The model (1)–(12) requires thousands of constraints and continuous and 0–1 variables for real-life cases: for example, a problem with 18 part types, 14 machines, and 4 periods expands to a 0–1 model with 3691 constraints, 3385 continuous and 0–1 variables, 12 537 elements, and 1295 0–1 variables.

Solving this problem until optimality exceeds the tolerable time, even when using mainframes and state-of-the-art optimizers such as OSL/VM (Wilson *et al.*, 1992). It still takes around 6 minutes of CPU time to obtain the first integer solution on an IBM 3090-400E.

Given the large-scale nature of the problem, it is unrealistic to seek for the optimal solution in our environment. Instead of providing optimal solutions, we favor a decomposition method that does not consider the integrality of all 0–1 variables at once, but considers them successively. Basically, it consists of analyzing the characteristics of the instance and considering iteratively the integrality of some subset of variables. Here, the first iteration optimizes the original model, where the integrality constraint of all variables is relaxed, except the 0–1 variables that belong to the first time period. After solving the subproblem associated with a given iteration, the 0–1 variables relating to the given period will be permanently fixed to the optimal solution for the subproblem except when an infeasibility occurs in later periods. The optimal solution of the last subproblem should lead to a near-optimal solution of the original problem. Elsewhere (Dillenberger *et al.*, 1994) we report an extensive computational experience for some real-life problems.

Finally, let us compare the period-by-period model with the global model (see Table 8.4). The average size of the submodel solved per period (especially the number of 0–1 variables) is significantly smaller than the size of the global model. This results in 93.8 seconds of CPU time for all four submodel optimizations, compared with 362.3 seconds for the global model. In both approaches, the optimization was stopped when the first integer solution was found. The LP solution for the global model is 126 264.1. It is a lower bound for the optimal solution. The optimality gap of the first integer solution for the global model is

Table 8.4 Model comparison.

	Global model	Period-by-period
Model size	3691 rows 3385 columns 12 537 elements 1295 0–1 variables 43 SOS sets	1562 rows 1798 columns Average 5368 elements 177 0–1 variables 0 SOS sets
LP solution	126 264.1	
1st MIP solution	128 496.6 after 219 nodes and 14 107 iterations	128 514.3 after 192 nodes and 4771 iterations
GAP	1.76%	1.78%
CPU time	362.3 seconds	93.8 seconds

only 1.76%. The optimality gap of the solution for the period-by-period approach is only 0.02% worse than the gap for the global model, but reached much faster. Especially for practical problems it is more important to provide near-optimal, but fast, solutions rather than optimal ones.

The period-by-period model has been implemented as the optimization kernel of an IBM internal decision-support system known as DACAPO (Demand Allocation for Capacity Optimization). This tool has been designed in such a manner that the model generator and the algorithm are completely transparent to the user. No knowledge of mathematical programming is required to use the tool.

The user interface allows parameter specification and start of the optimization run. It gives access to the results, too. The results can be represented either in graphical or numerical form. The parameters and results are stored in a relational database. Therefore it is easy to exchange data with other applications and to build user-defined reports. Depending on the parameters, the optimization kernel generates and solves the mathematical model using OSL as basic software.

REFERENCES

Dillenberger, Ch., Escudero, L. F., Wollensak, A., and Zhang, W. (1994) On practical resource allocation for production planning and scheduling with period overlapping setups, submitted to *European Journal of Operational Research*.

Fleischmann, B. (1988) Operations-Research-Modelle und -Verfahren in der Produktionsplanung, *ZfB*, **58** (3), 347–372.

Fromm, H., Dillenberger, Ch., and Wollensak, A. (1993) Optimization in microelectronics manufacturing, in: T. A. Ciriani, and R. C. Leachman (Eds.), *Optimization in Industry*, John Wiley & Sons, Chichester.

Tempelmeier, H., and Kuhn, H. (1992) OR-Modelle zur Planung flexibler Fertigungssysteme. Ein Überblick, *OR Spektrum*, **14** (4), 177–192.

Wilson, D. G., and Rudin, B. D. (1992) Introduction to the IBM Optimization Subroutine Library, *IBM Systems Journal*, **31** (1), 4–10.

Wollensak, A., Dillenberger, Ch., and Fromm, H. (1992) A decision support system for production planning and scheduling systems, *Journal of Decision Systems*, **1** (2–3), 261–272.

IBM German Manufacturing Technology Center
Dept 0817, Bldg 71065-53
PO Box 266
71044 Sindelfingen
Germany

9
Manufacturing–Retail Partnership through Global Logistics Optimization

Hua Thanh Binh
The HUA Group, Paris, France

ABSTRACT
In order to get access to the European Single Market we need a 'win–win' manufacturing–retail partnership, interconnecting individual and collective intelligence in the field of global logistics optimization. In such a fierce competitive environment large—and hitherto unexplored—sources of shared productivity, effectiveness and competitiveness may be exploited in 'extended enterprise' supply-chain management. This optimized collaborative dynamic management system is different from and more effective than both 'EDI'-type administrative partnerships and traditional confrontational relationships between sellers and buyers.

The HUA approach consists in combining, in a dynamic and flexible way, MRP/DRP centralized models and JIT/ICS (inventory control system) decentralized philosophies, to give an optimized solution to each individual node of the 'extended logistics network' (ELN) from the supply of raw materials to the points of sale delivering the end-consumers. HUA 5557 is the software tool of this individual intelligence, the core of which is described in this chapter.

EAGLE (European Advanced Global Logistics Enterprise) is a European strategic project (under ESPRIT III) which has as an objective the building of the basis of a new generation of decisions support system (DSS) exploiting advanced IT to offer fast, intelligent and flexible response interconnecting individuals, local intelligence, overall culture, and the behavior of each actor in a global multi-site multi-supplier logistics coordination of the whole ELN.

The technical base of the HUA 'multimodel' for individual intelligence needs intensive computing power at the local sites (up to 2 million calculations for just one optimized supply order); and like the flexible EAGLE multi-site multi-supplier global logistics coordination, a new type of manufacturing retailing collaborative intelligence, this is technically feasible. A demonstration of this feasibility was made at CeBIT '93 in Hanover, Germany.

Optimization in Industry 2. Edited by T. A. Ciriani and R. C. Leachman

1 INTRODUCTION

1.1 Objective: The Company's Survival

The completion of the European Single Market, fierce competition, high inventory costs, shorter product life cycles, demands for Just-in-Time delivery, ever improving customer service and the need for anti-waste resources management have all led manufacturers and retailers to reexamine their entire commercial, logistics, and manufacturing operations. Competitive advantage and survival are increasingly dependent on the ability of each actor to cost-effectively deliver the right quantity of the right product, at the right place and time, and in prime condition.

1.2 Weaknesses of the Traditional Approaches

In place of 'wait and see', 'fatalist' and 'opportunistic' behaviors, the 'active' approaches for a company's survival may be classified in two main categories: the 'political' and the 'Information technology' (IT) solutions.

The 'political' solutions are mainly based on capitalist and strategic confrontational and/or 'cartel' relationships between selling and buying powers; these solutions are generally dubious. Among the different classes of IT solutions, we may distinguish three types of approach—physical, information, and intelligence—following response-flexibility level codes 1–5 described in Figure 9.1.

Making the physical tool flexible as a top priority—often referred to as CIM (computer-integrated management) at the shop-floor level—is also a dubious solution and had some spectacular failures in the 1980s.

The most traditional IT solutions consist in the ability of response-flexibility level 2 to rationalize and normalize the company's information system for better intra- and inter-enterprise communication. They include EDI tools and new manufacturing and material management design. They are generally costly, with uncertain ROI (return on investment), and are more efficient than effective as basically an informational administrative and deterministic approach.

There are no significant existing solutions in the 'intelligence' level. On levels 3 and 4 (operational and tactical levels) most traditional IT tools are deterministic MRP/DRP-based; on level 5 (strategic level), tools are deterministic OR-based, or are parts of an EIS (executive information system) without any added intelligence.

None of these traditional IT solutions offers a good tool for the manufacturing–retailing 'win–win' collaborative management on the whole ELN which is a *sine qua non* condition for the company's survival in the 1990s. The HUA/EAGLE 'intelligence' IT solution described in this chapter may be an alternative for a new type of management.

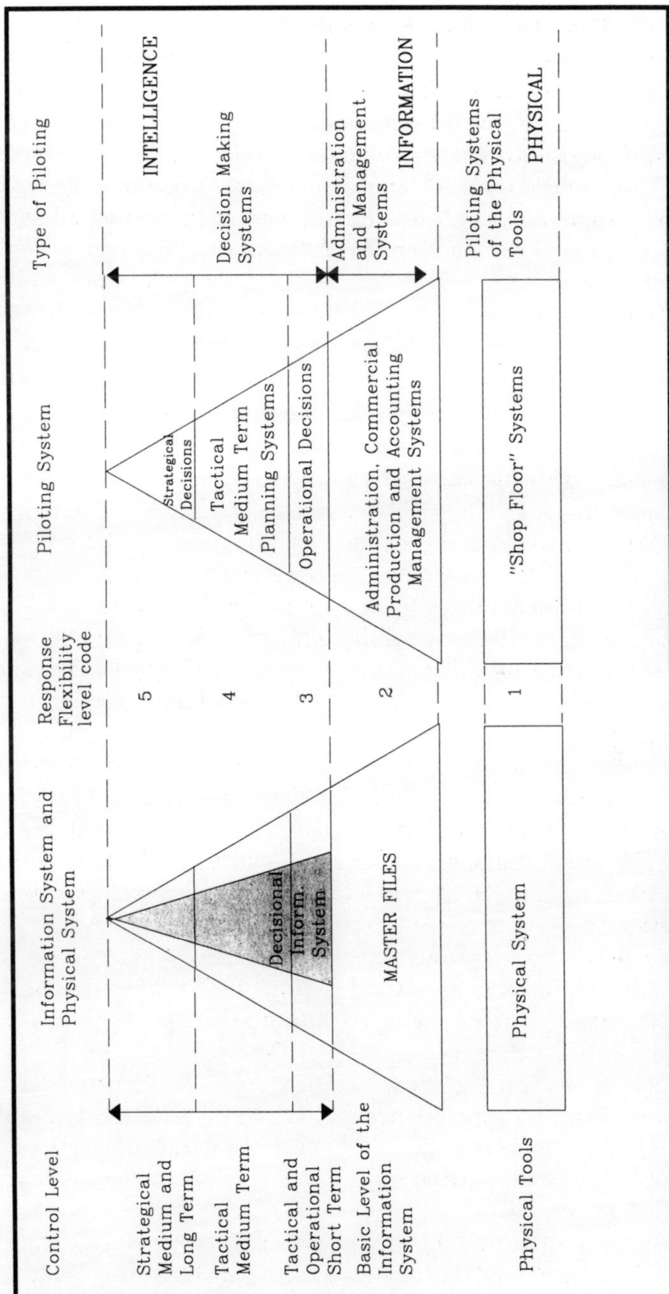

Control Level
Information System and Physical System
Response Flexibility level code
Piloting System
Type of Piloting
Strategical Medium and Long Term
Tactical Medium Term
Tactical and Operational Short Term
Basic Level of the Information System
Physical Tools
Decisional Inform. System
MASTER FILES
Physical System
5
4
3
2
1
Strategical Decisions
Tactical Medium Term Planning Systems
Operational Decisions
Administration, Commercial Production and Accounting Management Systems
"Shop Floor" Systems
INTELLIGENCE
Decision Making Systems
Administration and Management Systems
INFORMATION
Piloting Systems of the Physical Tools
PHYSICAL

Figure 9.1

2 TECHNOLOGICAL BASIS OF THE MANUFACTURING-RETAIL PARTNERSHIP

Computer-integrated management of the global logistics of the extended enterprise may be studied on four different levels of control and integration:

(i) local site of the manufacturing or retailing industry;
(ii) logistic chain of the retailer or of a multi-site manufacturing company;
(iii) global logistic enterprise (GLE): integration of the separate manufacturing and retailing activities (ii) and with agreed information being passed across the partnership interface, for management and dynamic control purposes;
(iv) global logistic extended enterprise (GLEE): complete integration at all levels and complete collaboration and information access.

This is illustrated in Figure 9.2 (see EAGLE Consortium, 1992a, b).

The first published results of the EAGLE/ESPRIT project 6599, as deliverables for CEC (Commission of the European Communities) in October 1992 were the following.

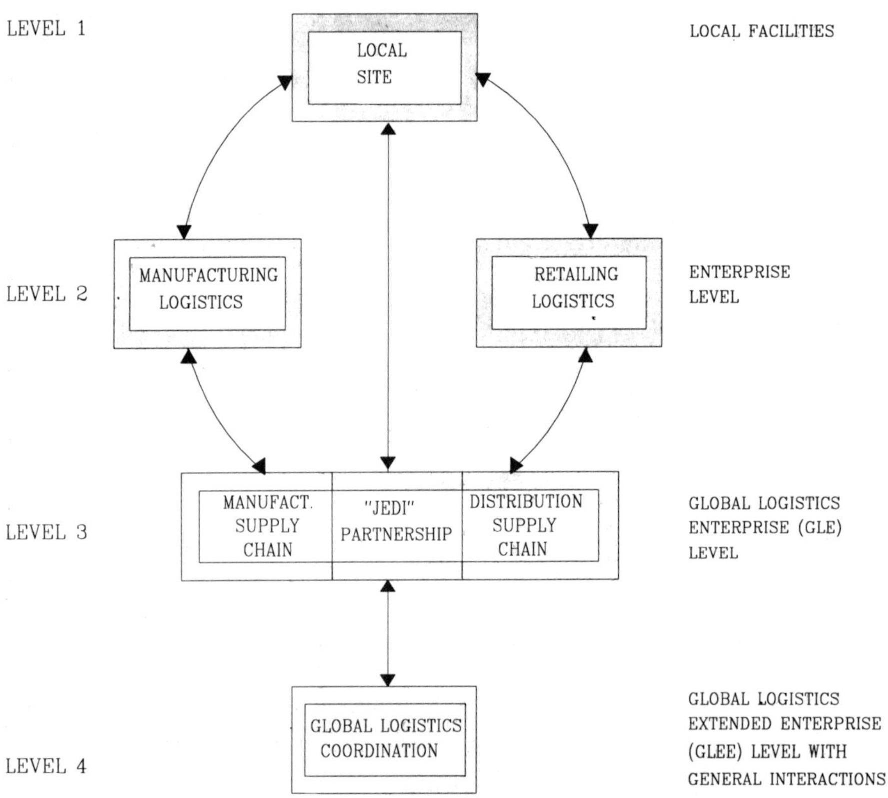

Figure 9.2

Control at the level of the local sites

On the whole ELN, for each local site (elementary nodes called 'PCR', points of control and regulation of the material flows), from raw-materials supply through manufacturing and distribution warehousing to transhipment platforms and points of sale, the HUA 5557 software is used as a leading-edge decentralized control tool. Its core is a triple interconnected autoadaptative multi-model (sales forecasting, stock-control optimization and supply-orders optimization as described below). This control of each local site is completely disconnected from its upstreams (suppliers) and its downstreams (customers). In contrast to the MRP/DRP 'push–flow' (from manufacturer to consumer) tunnel and the JIT 'pull–flow' (consumer-demand driven) sequential chain, it has the advantage of a fully market-driven approach.

Control of single-enterprise multi-site multi-supplier logistics

At this level, knowing in detail the actual and forecasted needs of each local-site 'PCR' for the right product at the right quantity and at the right time and place (from the HUA local control), it is easy to use traditional OR methods and modern methods such as CLP (constraint logic programming) to optimize the global logistics costs of a given organized—and chained—logistics network of PCRs (called an 'ECR', ensemble de contrôle et de régulation) from the same company, and *ipso facto*, with common interests. Hua's Herakles control ensures, thanks to the HUA 5557 bottom-up multi-PCR instructions and to the top-down autoadapted guidelines of the forecasted master program schedule of the ECR, a global logistics optimization taking into account in real time the multiple local constraints, the target, and not always 'rational' behaviors.

Multi-site multi-supplier multi-enterprise global logistics control

Current approaches to solving the problems of control at this level have led to impractical model formulations (with mega-matrices of 5 million variables!), due to the exponential increase in constraint complexity and the use of unsuitable model structures.

The modeling under EAGLE/ESPRIT III offers a smart and usable bottom-up/top-down solution between organized single-enterprise ECRs, interacting with their PCRs acting as individual actors, in the framework of the free-market economy, with theorically 'pure and perfect competition' and 'transparency of the information system', etc.

Under these technical conditions, a flexible IT control of the 'rules of the game' may be piloted in a 'JEDI' (just enough desirable inventory) partnership, which is 'win–win' for manufacturers and retailers, or in confrontational relationships, generating disturbing promotion buyings of wholesalers against manufacturers. EAGLE's control of an SCR (system of control and regulation, organized set of ECRs), will add macro-connectionism, distributed neural artificial intelligence, for processing scenarios with incomplete information, fuzzy logic as well as management compromises and justifications of complex decisions involving individual and collective interests. (This is the target of EAGLE 2.)

General coordination control

The EAGLE modeling represents a significant step towards the complete integration of the detailed management of the logistics network. This part of the system will be developed in a private project by a European transregional global logistics network (PERCEVAAL).

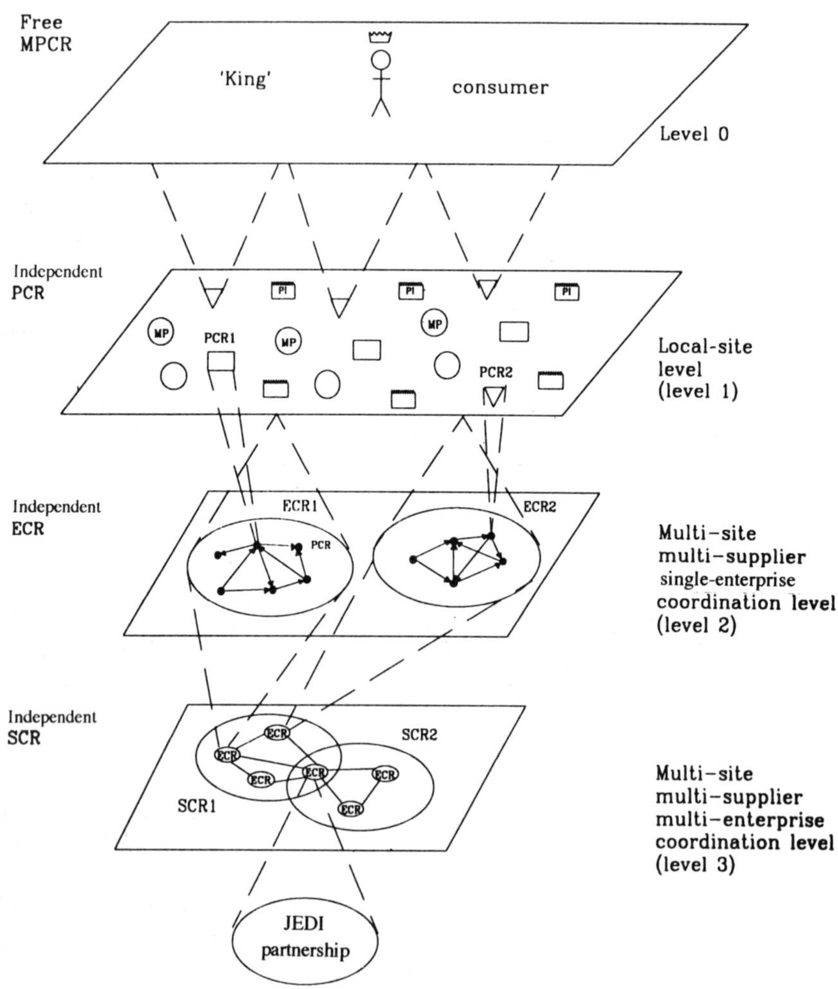

Figure 9.3

General presentation of the EAGLE model

Under the representation of Figure 9.3 we may find all possible traditional cases of 'global manufacturing' as simplistic and rigid subsets (Hua, 1992) as well as, on the conceptual level, the concept of 'Holons and Holarchy' (perhaps equivalent to PCRs/ECR?) proposed by the Japanese IMS productivist paradigm of the MITI for the year 2002.

3 HUA'S OPTIMIZATION SOLUTIONS IN INDUSTRY

As described in Section 2, the solution to this complex problem of controlling multi-site multi-enterprise logistics, provided by the EAGLE project, depends fundamentally on the reliability, robustness and power of HUA's optimization at the local-site level (PCR or elementary node of the ELN).

MRSHUA (Multimodèle HUA de Régulation par Servomécanisme) is a multi-model invented twenty years ago and has been the mathematical basis of the HUA 5557 software package (produced in 1982), running in more than 200 PCR local sites in Europe, concerning more than 100 different companies, large or SMEs (small and medium-sized enterprises), in manufacturing and retailing industries (Curth, 1992; *Point de Vente*, 1989; *Manutention*, 1985; Puget, 1990, 1991, 1992; *Logistique Magazine*, 1990). It is used by several of the top-ten largest European retailing groups. HUA 5557 provides a leading-edge optimization solution for dynamic market-driven material flows management, essentially on the PCR local-site level.

3.1 Main functions of the HUA 5557 DSS

For each management period, tactical macro-regulation planning, or operational reordering system, the main functions of the HUA system follow the classical decision-making process:

- analysis of the information,
- forecast from the information,
- optimization and simulation of decisions,
- validation and transmission of decisions.

3.2 Functions of local-site control

All along the global supply chain, ranging from the most downstream sites such as points of sale (PCR logistic-level 1) delivering to the end-consumers (terminal

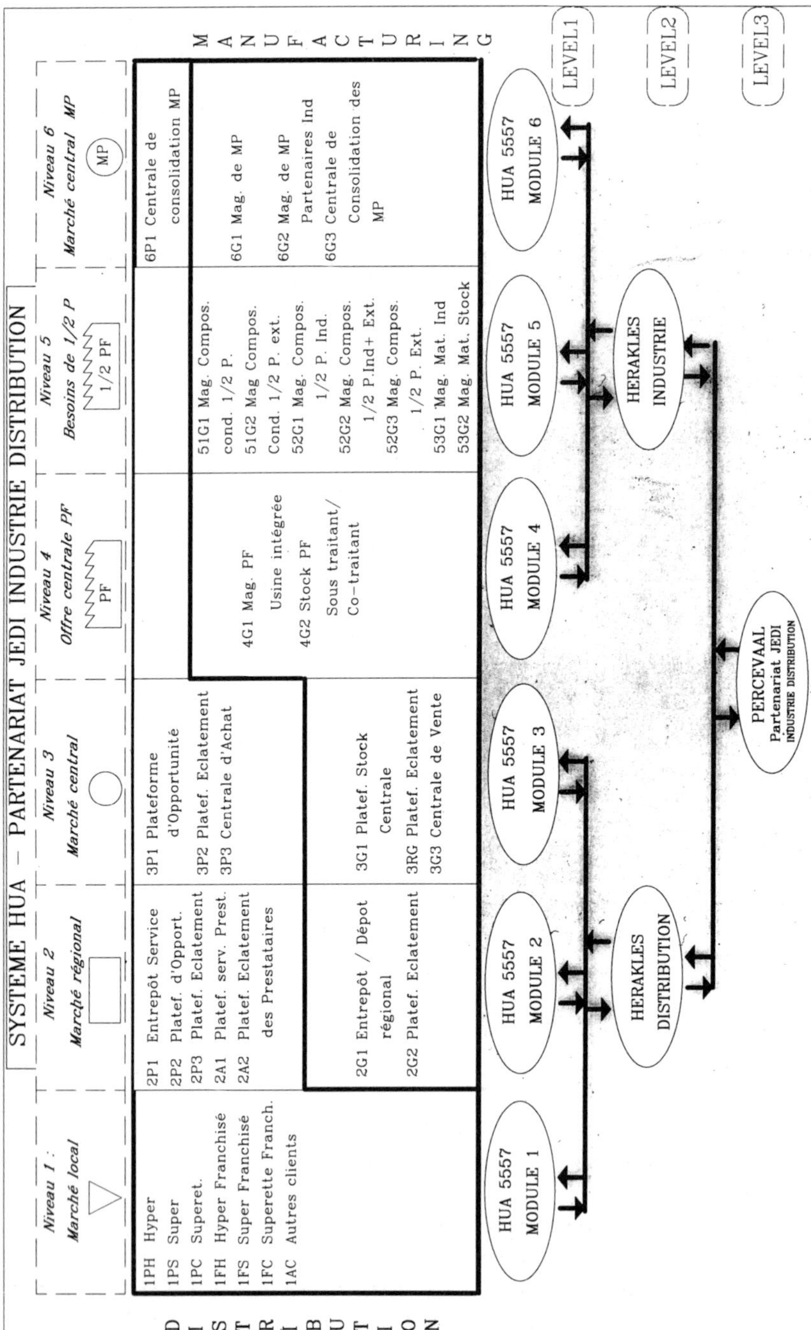
SYSTEME HUA – PARTENARIAT JEDI INDUSTRIE DISTRIBUTION
Niveau 1 : Marché local
Niveau 2 Marché régional
Niveau 3 Marché central
Niveau 4 Offre centrale PF
PF
Niveau 5 Besoins de 1/2 P
1/2 PF
Niveau 6 Marché central MP
MP
1PH Hyper
1PS Super
1PC Superet.
1FH Hyper Franchisé
1FS Super Franchisé
1FC Superette Franch.
1AC Autres clients
2P1 Entrepôt Service
2P2 Platef. d'Opport.
2P3 Platef. Eclatement
2A1 Platef. serv. Prest.
2A2 Platef. Eclatement des Prestataires
2G1 Entrepôt / Dépot régional
2G2 Platef. Eclatement
3P1 Plateforme d'Opportunité
3P2 Platef. Eclatement
3P3 Centrale d'Achat
3G1 Platef. Stock Centrale
3RG Platef. Eclatement
3G3 Centrale de Vente
4G1 Mag. PF Usine intégrée
4G2 Stock PF Sous traitant/ Co-traitant
51G1 Mag. Compos. cond. 1/2 P.
51G2 Mag Compos. Cond. 1/2 P. ext.
52G1 Mag. Compos. 1/2 P. Ind.
52G2 Mag. Compos. 1/2 P.Ind+ Ext.
52G3 Mag. Compos. 1/2 P. Ext.
53G1 Mag. Mat. Ind
53G2 Mag. Mat. Stock
6P1 Centrale de consolidation MP
6G1 Mag. de MP
6G2 Mag. de MP Partenaires Ind
6G3 Centrale de Consolidation des MP
DISTRIBUTION
MANUFACTURING
HUA 5557 MODULE 1
HUA 5557 MODULE 2
HUA 5557 MODULE 3
HUA 5557 MODULE 4
HUA 5557 MODULE 5
HUA 5557 MODULE 6
HERAKLES DISTRIBUTION
HERAKLES INDUSTRIE
PERCEVAAL Partenariat JEDI INDUSTRIE DISTRIBUTION
LEVEL1
LEVEL2
LEVEL3

Figure 9.4

flow, logistic-level zero), to the most upstream sites such as raw-materials stores (PCR logistic-level 6), HUA's single-site control has the same functions of analysis, forecast of sales or demand, scheduling of JIT stock requirements, order optimization and simulation, issuing orders (by telex, fax, EDI, etc) as described in the last subsection.

These functions are presented, as module 1 to 6 (the same numbers as PCR logistic level) in Figure 9.4.

3.3 Functions of multi-site multi-supplier single-enterprise control

For each organized set of PCRs linked with an ECR supply-chain management, the Herakles control consists mainly in coordinating, on the one hand, the macro-regulation planning (as in the MRP/DRP approach, but just with flexible guidelines) and, on the other hand, the interactions between PCRs on orders and sales forecast and autoadaptation with the ECR's guidelines. These functions are presented in Figure 9.4.

4 HUA'S OPTIMIZATION CORE

The core of the HUA system is compound with a triple interconnected and autoadapted multi-model: (A) sale or demand forecast, (B) dynamic scheduling of the JIT stocks requirement, (C) JEDI order optimization (EAGLE Consortium, 1992a, b). We present hereafter the algorithms (guide lines) of multi-models B and C and the local global interaction.

The multi-model B of the MRSHUA, downstream to the demand forecasts and upstream to the orders optimization, plays a fundamental role in the calculation of the 'minimal rules of the game' for the equilibrium of the elementary market represented by the PCR:

$\hat{X}_t^a$ demand forecasts of the article *a* at time *t* (hour, day, week, month),
σ_1^a smoothed standard deviation on forecasts/results autoadapted at the period *t*,
K^{a,C_S} security stock coefficient of the article *a* depending on the requested quality of service C_S for this article on the PCR,
σ_2^a delay deviation (in days worked) of the supplier *F* for the article *a*,
$\overline{X}_a$ daily average consumption of the article *a*,
p periodicity (in hours, days, weeks, or months),
d average delay of the supplier (in hours, days, weeks, or months),
S_t^a stock of the article *a* at the beginning of the period *t*,
EC_t^a order pool to the supplier of the article *a* available in stock on the PCR at the beginning of the period *t*,
BL_t^a minimum stock requirement of the article *a* at the latest time *t*.

For all articles a of the supplier F:
Iteration for $t = 1$ to h

$$\mathrm{BL}_t^a = \sum_{k=t}^{t+p+d-1} \hat{X}_k^a + K^{a,t_s}\,(\sigma_1^a \sqrt{(p+d)} + \sigma_2^a / 2\ X^a) - S_t^a - \sum_{k=t}^{t+p+d-1} \mathrm{EC}_k^a,$$

$$\mathrm{BL}_t^a = \mathrm{Max}\ (0, \mathrm{BL}_t^a),$$

$$\mathrm{EC}_{t+d}^a = \mathrm{EC}_{t+d}^a + \mathrm{BL}_t^a,$$

$$S_{t+1}^a = S_t^a - \hat{X}_t^a + \mathrm{EC}_t^a.$$

p_u^a total unit price of the article a ex-works (given the ordered quantities and the promotion periods),
p_l^a logistics costs linked to the article a given the transport hierarchical price scale,
CL^F logistics costs for the whole of the article a ordered by the supplier F given the total ordered quantities,
coefQ^a volume conditions coefficient of the article a,
Q_W^F imposed volume conditions on the order to the supplier F.

We have three groups of constraints:

(C1) minimum-flow equilibrium constraints,

(C2) logistics, purchasing, pricing, and technical constraints,

(C3) speculative purchase constraints.

The multimodel C of MRSHUA consists in determining for each supplier F, the supply orders (OA) matrix fixed (at the time 1) and forecasted (from the time 2) which gives the following optimized matrix:

$(\hat{\mathrm{OA}}_t^a)_{\text{for } t=1 \text{ to } h}^{\text{all } a\in F}$ which realizes:

$$\mathrm{Min}\ \left(\sum_{\text{for all } a\in F} (\mathrm{OA}_t^a(p_u^a + p_l^a)) + \mathrm{CL}^F\right)$$

under the conditions

$$\sum_{\text{for all } a\in F} \mathrm{OA}_t^a \times \mathrm{coefQ}^a \geqslant Q_W^F$$

and the constraints (C1), (C2), and (C3).

If all conditions and constraints are realized, we will have the 'classical' automatic chained sequential transmission shown in Figure 9.5. Otherwise, the HUA system proposes two alternatives:

- automatic re-looping of the information ('what if' or 'how to achieve' simulations) as presented in Figure 9.6;
- human or external autoadaptative interactions as presented in Figure 9.7.

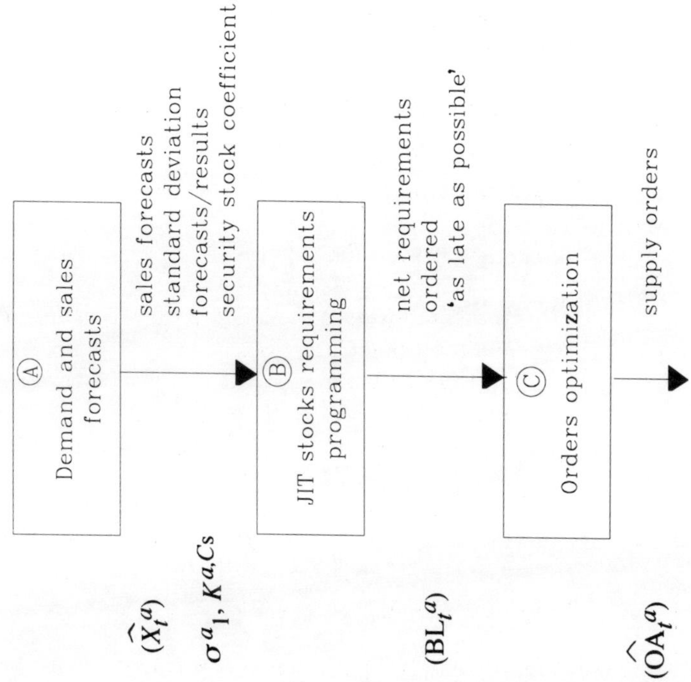

Figure 9.5

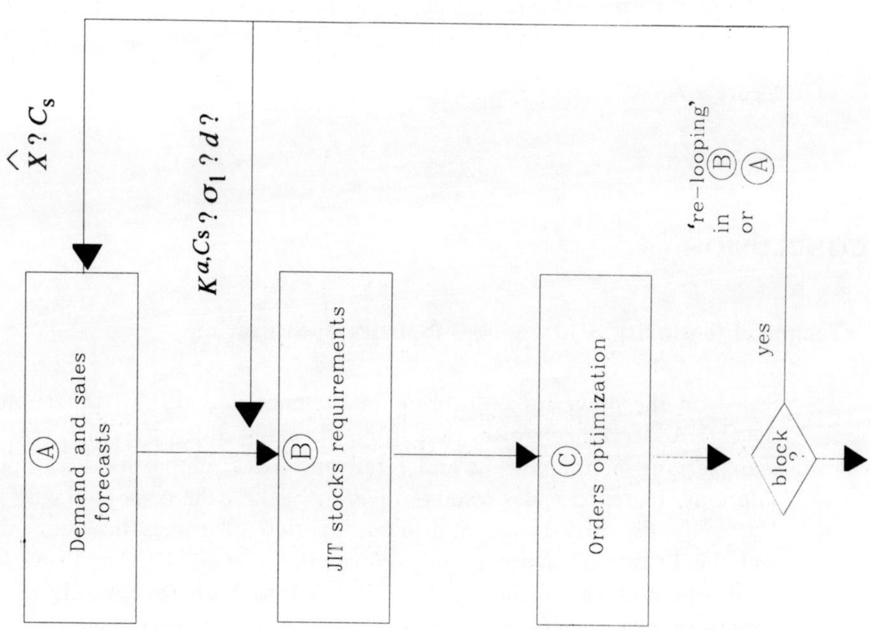

Figure 9.6

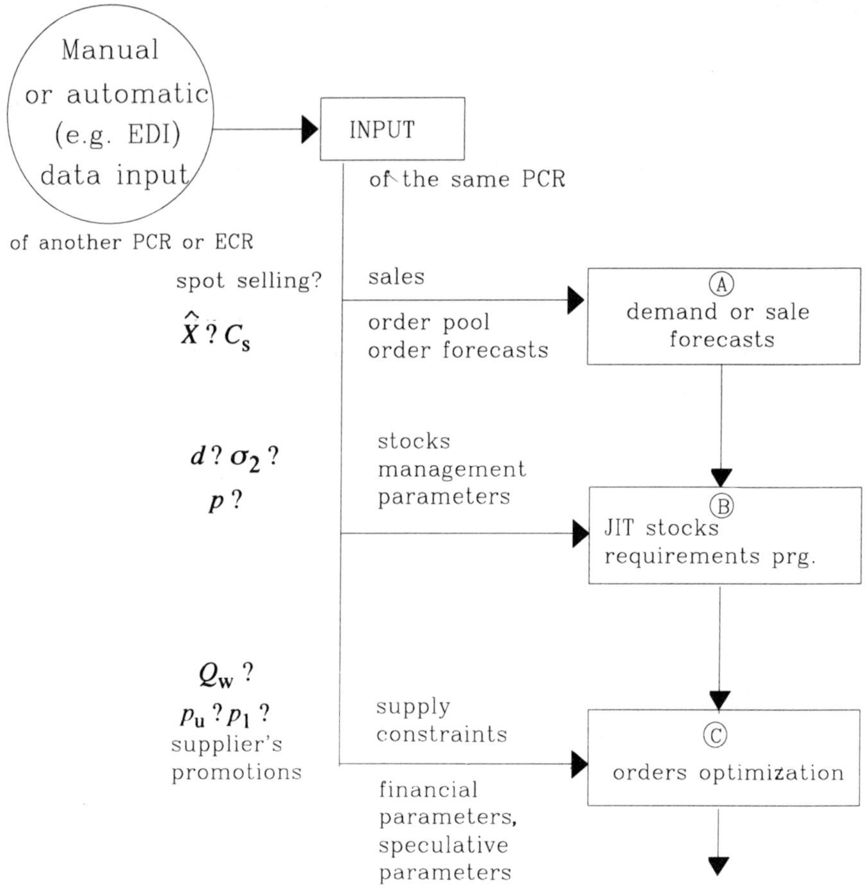

Figure 9.7

5 CONCLUSION

5.1 Technical feasibility of the global logistics optimization

Based on the powerful bottom-up interconnections of individual intelligences, the HUA servomechanism has offered, daily over the last ten years, to managers in manufacturing and retail industries, global logistics optimization solutions. The spectacular results—dividing by three the stock level with practically 100% quality of service, multipling by two or three the sales, the profits and the ROI, demonstrate the performance of the HUA multi-model on the local site and the technical feasibility of the whole EAGLE concept and architecture.

5.2 Advantages of the global-logistics JEDI partnership

Traditional configurations of purchasing powers acting against selling powers will not achieve the European Single Market access for most of SMEs. The JEDI partnership, being based on manufacturing–retail global-logistics optimization, is a 'win–win' type of solution which may involve, on just the financial level, an efficiency multiplied by a factor of 10 for the user companies. This enhancement could be the same order of magnitude as the one brought about by the lean-manufacturing approach to mass car production in Japan over the last 20 to 30 years.

The European advanced IT–intelligence solution is a worldwide leading-edge technology bypassing all traditional approaches like MRP/DRP or pure JIT in the general field of global manufacturing management of the extended enterprise.

REFERENCES

Curth, M. (1992) Antwort von der Pyramide, *Harvard Manager* No. 3, 129–135.

EAGLE Consortium (1992a) *ESPRIT Project 6599, EC deliverable D112A: Initial report on the enterprise modelling of the extended logistics networks*, Hua T. B., HUA-ECG, Paris, November.

EAGLE Consortium (1992b) *ESPRIT Project 6599, EC deliverable D12A: Initial report on the development of Decisions Support Systems algorithms*, Hua T. B., HUA-ECG, Paris, November.

Logistique Magazine (1990) Stocks: Privilégier le spéculatif, *Logistique Magazine*, No. 45, February.

Manutention (1985) Pour une gestion industrielle optimisée, *Manutention*, No. 87, 50–51, 25 February.

Point de Vente (1989) La logistique joue la continuité, *Point de Vente*, No. 371, 36–38, 1 October.

Puget, Y. (1990) Intermarché pilote ses achats, *LSA*, No. 1192, 12 January.

Puget, Y. (1991) Progiciel HUA 5557: le spéculatif gagne la mise, *LSA*, No. 1254, 25 April.

Puget, Y. (1992) Le partenariat assisté par ordinateur, *LSA*, 23 April.

Chairman of the HUA Group
21 Boulevard Delessert
75016 Paris
France

10
Project Scheduling with Parallel Activities, Time-window and Precedence Constraints

Anna Sciomachen
University of Milan, Italy

ABSTRACT
A methodology for sequencing and scheduling the activities to be performed in a project is presented, so as to allow activities to overlap and independent execution of some sequences of activities. This methodology also allows precedence relationships to be enforced as well as time-window constraints, defined by the release date and deadline associated with each activity. Given the duration of the activities and the setup between activities, the objective is to find a feasible sequencing of all the activities such that the completion time of the project is minimized. A 0–1 model of the problem is presented along with an application to a robot assembly line.

1 INTRODUCTION AND PROBLEM DEFINITION

The problem considered in this chapter is an extension of the so-called sequential ordering problem (SOP) (Escudero and Sciomachen, 1990) in which a set of jobs has to be processed on a single machine and time availabilities (defined by deadlines and release dates associated with the jobs) and precedence constraints between jobs are given. The SOP is formulated on a direct graph $G=(V,A)$ with node set V, of size n, and arc set A, of size m, with weights on the nodes (jobs) and on the arcs, and with restrictions on the weights of subpaths from the origin to certain specified nodes. These constraints require that the weight of a subpath from the first node to a specific node be no less than a minimum value (release date) and no more than a given maximum value (deadline). The objective of the SOP is to find a minimum-weight Hamiltonian path in G so as to satisfy precedence constraints between nodes, given by the arc set P in a direct graph $R=(N,P)$, for $N \subseteq V$.

Optimization in Industry 2. Edited by T. A. Ciriani and R. C. Leachman

The extension presented in this chapter considers the possibility of jobs overlapping and also independent execution of some sequences of jobs, i.e. splitting the Hamiltonian path into different sequences consisting of independent elementary paths. The main application area lies in machine sequencing for flexible manufacturing systems where no more than one operation can be processed simultaneously by a single machine and sequences of operations of the same production plan can be performed in parallel. (For the mathematical aspects of scheduling theory and its applications see Coffman, 1976, and Bellman *et al.*, 1982.)

Formally, consider the digraphs $G=(V,A)$ and $R=(N,P)$. A weight p_i and the attributes r_i and d_i are associated with each job (node) $i \in V$, a weight c_{ij} is associated with each arc $(i,j) \in A$, and a weight b_{ij} is associated with each arc $(i,j) \in P$. Here p_i represents the processing time of job i, r_i is the release date of job i (i.e. the earliest time at which job i can start to be processed), d_i is the deadline of job i (i.e. the latest time at which job i can be completed), and c_{ij} represents the setup time, i.e. the time required by a machine for changing from job i to job j. The relationship $(i,j) \in P$ for $b_{ij} \geqslant 0$ means that node i must be sequenced before node j with a minimum lag in the ordering of the nodes. The relationship $(i,j) \in P$ for $b_{ij} < 0$ means that node j must be sequenced after node i with a maximum overlap of $|b_{ij}|$ time units. Let us suppose that $p_i, r_i, d_i \in Z^+$ $\forall i \in V$, $c_{ij} \in Z$ $\forall (i,j) \in A$, $b_{ij} \in Z$ $\forall (i,j) \in P$, i.e. we do not restrict the weights on the arcs A and P to be nonnegative.

The problem consists of finding a set of elementary paths in G such that the time window (defined by r_i and d_i) for scheduling the execution of each node is satisfied; the precedence constraints and maximum overlapping between nodes, given by the relationships defined by the arc set P, are satisfied; and the makespan, i.e. the value of the objective function at the node sequenced last, is minimized. The objective function, representing the completion time of the job processed last, is defined in terms of p_i, r_i, c_{ij} and b_{ij}, $\forall i \in V$, $\forall (i,j) \in H$, where H is a path in any feasible solution. In particular, let t_j be the completion time of job $j, j \in V$, in any path H. If i is the job processed immediately before j in H then t_j is given by:

$$t_j = \max\{r_j,\ t_i + c_{ij},\ t_k + b_{kj}\ \forall k \in V \mid (k,j) \in P\} + p_j. \tag{1}$$

Note that the number of elementary paths of any feasible solution corresponds to the number of machines involved in the processing of all the jobs in V, and each path gives the ordering of the jobs processed by a single machine. Moreover, note that the relationship between two given nodes, say i and j, depends on the relative values b_{ij} and c_{ij}. In particular, the following relationships hold.

$(i,j) \in A$ with $c_{ij} \geqslant 0$ represents the possibility of sequencing job j immediately after job i in the same path, provided that the following condition is satisfied:

$$t_j - p_j - t_i \geqslant c_{ij}. \tag{2}$$

$(i,j) \in A$ with $c_{ij} < 0$ represents the possibility of sequencing job j immediately after job i in the same path and such that the execution of jobs i and j can be overlapped by at most $|c_{ij}|$ time units; condition (2) must be satisfied.

$(i,j) \in P$ with $b_{ij} \geqslant 0$ forces job j to be sequenced after (immediately or not) job i, either in the same path or in different paths, and such that between the completion of job i and the beginning of job j at least b_{ij} time units must have elapsed, independently of how many jobs are executed in between, i.e. the following condition must be satisfied:

$$t_j - p_j - t_i \geqslant b_{ij}. \tag{3}$$

$(i,j) \in P$ with $b_{ij} < 0$ represents the possibility of processing job j either before job i or in parallel (in a different path) to job i, provided that between i and j there is a maximum overlapping of $|b_{ij}|$ time units, i.e. that condition (3) is satisfied.

Let us now present all four possible relative orderings, in the same path or in different paths, between two nodes $i,j \in V$, in any feasible solution of the problem.

(a) $c_{ij} \geqslant 0$ and $b_{ij} \geqslant 0$. If i and j are sequenced in the same path then job j can be processed immediately after job i provided that condition (4) is satisfied. Otherwise, if i and j belong to different paths, then condition (3) must be satisfied.

$$t_j - p_j - t_i \geqslant \max\{c_{ij}, b_{ij}\}. \tag{4}$$

(b) $c_{ij} \geqslant 0$ and $b_{ij} < 0$. If i and j are sequenced in the same path then job j can be processed immediately after job i provided that condition (2) is satisfied. Otherwise, if i and j belong to different paths, then condition (3) must be satisfied.
(c) $c_{ij} < 0$ and $b_{ij} \geqslant 0$. Independently of whether or not i and j are in the same path, condition (3) must be satisfied.
(d) $c_{ij} < 0$ and $b_{ij} < 0$. If i and j are sequenced in the same path then jobs i and j can be overlapped, provided that condition (5) is satisfied. Otherwise, if i and j belong to different paths, then condition (3) must be satisfied.

$$t_j - p_j - t_i \geqslant \min\{c_{ij}, b_{ij}\}. \tag{5}$$

Note that for a given pair of jobs $i,j \in V$, if $(i,j) \in A$ and $(i,j) \notin P$ there is no relationship constraint between i and j, so that i and j can be processed either in different paths or in the same one, provided that if i and j are sequenced in the same path H then j can be processed immediately after i only if condition (2) is satisfied according to the value of c_{ij}; if $(i,j) \notin A$ then necessarily job j cannot be processed immediately after job i.

Note that if $b_{ij} = 0 \ \forall (i,j) \in P$ then the problem reduces to the SOP. Moreover, if $P = \emptyset$ and no release dates and deadlines are given then the problem reduces to the asymmetric traveling salesman problem (ATSP) (Lawler *et al.*, 1985). Unfortunately, at least to the author's knowledge, there has been no previous work on the extension of the ATSP in the direction of the problem presented here (see, e.g. Wittrock, 1986, 1987; Ahmadi *et al.*, 1992).

In the next section a preprocessing phase for reducing the size of the problem and tightening the constraints is presented. In Section 3 a 0–1 optimization model is given, while in Section 4 the model and the related solution are presented with respect to a case-study of a robot assembly line. Finally, some conclusions are given.

2 PREPROCESSING OF THE DIGRAPHS *G* AND *R*

Before looking for a feasible solution, a preprocessing phase for reducing the size of the problem and tightening the constraints is performed. In particular, in this phase we try to reduce as much as possible the gap between the deadline and the release date of each job, and to remove redundant arcs from both A and P.

In order to tighten the time windows, two main considerations can be made. First, in any feasible solution the processing of a given job, say j, cannot start before the completion of all the jobs that are predecessors of j in P and of the job, say i, immediately before j in the same path H. Formally, we can increase the value of the release date of each job $j \in V$ according to the following rule:

$$r_j = \max\{r_j, R_{1j}, R_{2j}\} \ \forall j \in V \tag{6}$$

where

$$R_{1j} = \max\{r_i + p_i + b_{ij} \ \forall i \,|\, (i,j) \in P\},$$

$$R_{2j} = \min\{r_i + p_i + c_{ij} \ \forall i \,|\, (i,j) \in A\}.$$

Second, in any feasible solution a given job, say i, cannot be completed after the beginning of the processing of all the jobs that are successors of i in P and after the beginning of the job, say j, sequenced immediately after i in the same path H. Formally, we can reduce the deadline of each job $i \in V$ according to the rule:

$$d_i = \min\{d_i, D_{1i}, D_{2i}\} \ \forall i \in P \tag{7}$$

where

$$D_{1i} = \min\{d_j - p_j - b_{ij} \ \forall j \,|\, (i,j) \in P\},$$

$$D_{2i} = \max\{d_j - p_j - c_{ij} \ \forall j \,|\, (i,j) \in A\}.$$

The next step in the preprocessing phase is to delete redundant arcs from A. In particular, we can remove an arc $(i,j) \in A$ whenever we know that job j cannot be processed immediately after job i, i.e. that arc (i,j) cannot be in any feasible H. In this way, we can say that $(i,j) \notin A$ if any of the following conditions is satisfied:

$$r_i + p_i + c_{ij} + p_j > d_j, \tag{8}$$

$$(j,i) \in P, \tag{9}$$

$$(j,k),(k,i) \in P. \tag{10}$$

Condition (8) implies that job j cannot be processed immediately after job i in any feasible sequence due to the violation of its deadline. Condition (9) implies that job j cannot be processed immediately after job i in any feasible sequence since there is a precedence constraint between j and i. Note that if $(j,i) \in P$ job i must be sequenced after job j either in the same path or in a different one, independently of the value of b_{ji}. Condition (10) implies that job j cannot be processed immediately after job i in any feasible sequence since there is an implicit precedence constraint between j and i, independently of the values of b_{jk} and b_{ki}.

The last step in the preprocessing phase is to delete redundant precedence constraints. In particular, an arc (i,j) can be removed from P if the following condition is satisfied:

$$d_i + b_{ij} \leqslant r_j. \tag{11}$$

In fact, the precedence constraint between i and j is redundant whenever job i has to be completed before the time at which job j is released.

3 THE OPTIMIZATION MODEL

In this section the proposed 0–1 model for formulating and consequently solving the problem is presented. Having in mind the definition of the problem in terms of the digraphs G and R given in Section 1, let x_{ij} be our m 0–1 variables, such that, for each arc $(i,j) \in A$, $x_{ij} = 1$ means that arc (i,j) belongs to a path H in the optimal feasible solution and $x_{ij} = 0$ otherwise.

Before formulating the problem, let us consider two dummy nodes, namely 0 and $n+1$, to be added to V, such that $r_0 = p_0 = d_0 = 0$, and $p_{n+1} = 0$, $r_{n+1} = \max\{r_v + p_v \ \forall v \in V\}$, $d_{n+1} = \max\{d_v \ \forall v \in V\}$. Node 0 is connected with all the other nodes in V, i.e. $(0,i) \in A \ \forall i \in V \setminus n+1$, with weight $c_{0i} = S_i$ (S_i represents the setup time required for processing job i as a first job in the sequence). Node $n+1$ is connected with all the other nodes in V, i.e. $(i,n+1) \in A \ \forall i \in V \setminus 0$, with weight $c_{i,n+1} = 0$. Thus $G = (V,A)$ is such that $|V| = n+2$ and $|A| = m+2n$. The introduction of the two dummy nodes 0 and $n+1$ allows us to consider node 0 as the first node in any path H in the solution of the problem and node $n+1$ as the last one; consequently, node $n+1$ is in any solution the last job to be processed and then our objective is the minimization of the completion time of job $n+1$, expressed as:

$$\min\{t_{n+1}\}. \tag{12}$$

The constraints of the problem are expressed by relations (13)–(22):

$$t_{n+1} \geqslant t_v \ \forall v \in V, \tag{13}$$

$$r_v + p_v \leqslant t_v \leqslant d_v \ \forall v \in V, \tag{14}$$

$$t_i + c_{ij} + p_j - t_j \leqslant (d_i + c_{ij} - r_j)(1 - x_{ij}) \ \forall (i,j) \in A, \tag{15}$$

$$t_k + b_{kj} + p_j - t_j \leqslant 0 \ \forall k \in V : (k,j) \in P, \tag{16}$$

$$t_i - p_i + p_j - t_j \leqslant (d_i - p_i - r_j)(1 - x_{ij}) \ \forall (i,j) \in A : c_{ij} < 0, \tag{17}$$

$$\sum_{i:(i,v) \in A} x_{iv} = 1 \forall v \in V \setminus n+1, \tag{18}$$

$$\sum_{j:(w,j) \in A} x_{wj} = 1 \ \forall w \in V \setminus 0, \tag{19}$$

$$\sum_{i,j:(i,j) \in A, j \neq n+1} x_{ij} = n, \tag{20}$$

$$\sum_{j \in V} x_{0j} = \sum_{i \in V} x_{in+1}, \tag{21}$$

$$u_i - u_j + n x_{ij} \leqslant n - 1 \ \forall i,j \in \{1,2, \ldots, n\}: (i,j) \in A, \tag{22}$$

where $x_{ij} \in \{0,1\}$, $u_i \in R \ \forall i = 1, \ldots, n$, $t_i \in Z^+ \ \forall i = 0, \ldots, n+1$.

Note that the optimization model (12)–(22) consists of $m + 4n + 2$ decision variables.

The objective function (12) expresses the minimization of the makespan.

Constraints (13) force the completion time of job $n+1$ to be greater than the completion time of all the other jobs in V, i.e. they force node $n+1$ to be the last node in any feasible solution. Constraints (14) give a lower bound and an upper bound for the completion time of each job in V. Constraints (15) force the completion time of each job in V to include the setup time c_{ij} when $x_{ij} = 1$, i.e. when job j is processed immediately after job i. (Note that if $x_{ij} = 0$ relation (15) is always satisfied.) Constraints (16) represent the relationships given by the arc set P. Note that constraints (14)–(16) imply the definition of the completion time of any job given in (1). Constraints (17) are valid for $c_{ij} < 0$ and are active when $x_{ij} = 1$, forcing the processing of job j to start not before the beginning of the processing of job i. Constraints (18)–(20) are the usual assignment constraints. Constraints (21) imply that the number of successor nodes of node 0 is the same as the number of predecessor nodes of node $n+1$, i.e. they give the number of independent parallel machines (elementary paths of any feasible solution) that can be used for processing all the n jobs. Finally, constraints (22) are the subtour elimination constraints (SECs). (Note that nodes 0 and $n+1$, by definition, are not involved in the SECs.)

4 A CASE STUDY

In this section the present problem is described for a class of flexible assembly systems for robot beds. Robot assembly involves the assembly and joining of various electrical and electronic components to a robot bed. Automated assembly is accomplished using either robots or specialized equipment.

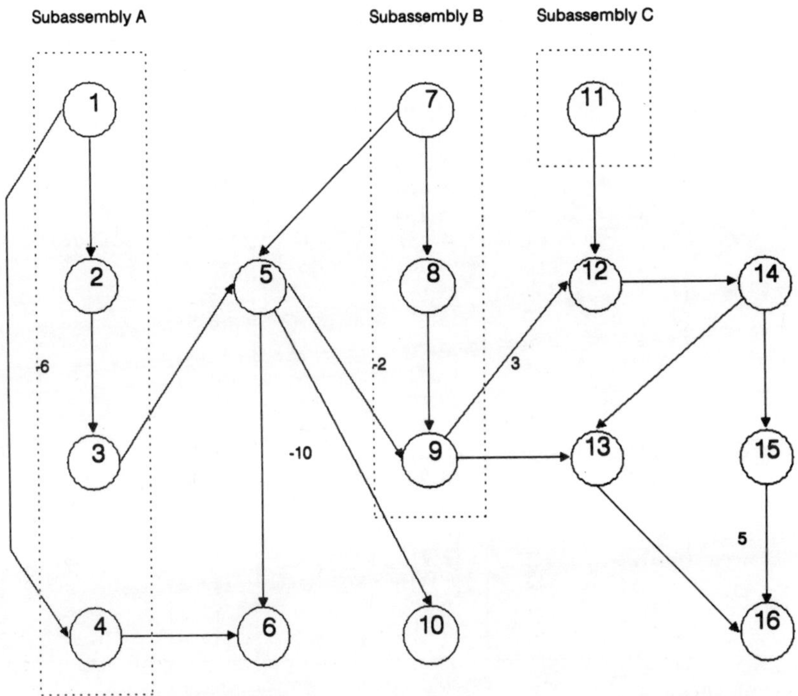

Figure 10.1 The relationships digraph *R*.

Today, the most common configuration for a robot assembly system is a set of independent assembly lines, such that each subassembly has a dedicated line. In the present case the assembly line consists of the following subassemblies (see Figure 10.1):

- subassembly A: defined by the bill of materials (BOM) and consists of operations 1, 2, 3, and 4;
- subassembly B: consists of operations 7, 8 and 9;
- subassembly C: consists of operation 11.

Note that the operations of different subassemblies can be completely overlapped and executed in parallel. In the proposed approach, operations that belong to independent subassemblies can be identified by a $-\infty$ setup, i.e. $c_{ij} = -\infty$ means that operations i and j can be executed independently.

The machines in the assembly lines need a setup that depends on the pair of operations to be performed consecutively. Some of the operations can be overlapped to some extent; conventionally, the absolute value for $c_{ij} < 0$ gives the maximum overlapping between the end of operation i and the beginning of operation j, provided that no operation is sequenced after operation i other than operation j. The setup times for each pair of operations (i,j) is reported in Table 10.1.

Table 10.1 Setup time c_{ij} for arc $(i,j) \in A$.

(i,j)	c_{ij}	(i,j)	c_{ij}	(i,j)	c_{ij}	(i,j)	c_{ij}
(1,2)	3	(4,10)	$-\infty$	(7,8)	1	(11,7)	$-\infty$
(1,4)	0	(4,11)	$-\infty$	(7,11)	$-\infty$	(11,8)	$-\infty$
(1,7)	∞	(4,12)	$-\infty$	(8,1)	7	(11,9)	$-\infty$
(1,8)	7	(4,13)	$-\infty$	(8,2)	7	(11,10)	$-\infty$
(1,11)	$-\infty$	(4,14)	$-\infty$	(8,3)	7	(11,12)	3
(2,3)	-6	4,15)	$-\infty$	(8,4)	$-\infty$	(12,4)	$-\infty$
(2,4)	0	(4,16)	$-\infty$	(8,5)	3	(12,6)	4
(2,7)	$-\infty$	(5,4)	3	(8,6)	3	(12,14)	2
(2,8)	7	(5,6)	1	(8,9)	-2	(13,4)	$-\infty$
(2,11)	$-\infty$	(5,8)	3	(8,10)	3	(13,6)	4
(3,4)	0	(5,9)	3	(8,11)	$-\infty$	(13,15)	0
(3,7)	$-\infty$	(5,10)	0	(9,4)	$-\infty$	(13,16)	2
(3,8)	7	(5,11)	$-\infty$	(9,6)	3	(14,4)	$-\infty$
(3,11)	$-\infty$	(6,8)	3	(9,11)	$-\infty$	(14,6)	2
(4,2)	-2	(6,9)	3	(9,12)	0	(14,15)	0
(4,3)	-2	(6,10)	3	(11,1)	$-\infty$	(15,4)	$-\infty$
(4,5)	3	(7,1)	$-\infty$	(11,2)	$-\infty$	(15,6)	2
(4,6)	1	(7,2)	$-\infty$	(11,3)	$-\infty$	(15,13)	0
(4,7)	$-\infty$	(7,3)	$-\infty$	(11,4)	$-\infty$	(15,16)	0
(4,8)	$-\infty$	(7,4)	$-\infty$	(11,5)	$-\infty$	(16,4)	$-\infty$
(4,9)	$-\infty$	(7,5)	2	(11,6)	$-\infty$	(16,6)	2

The weight b_{ij} denotes the smallest gap between the ending of operation i and the beginning of operation j, independently of what other operations are executed in between, if any. Note that $b_{ij}<0$ allows the overlapping of operations i and j (of at most $|b_{ij}|$ time units), either if operations i and j are executed by the same machine or if i and j are processed by different machines. Figure 10.1 gives the precedences graph $R=(N,P)$, where $|P|=19$, and then it gives the gap b_{ij} for $(i,j) \in P$.

It is worth mentioning that setup considerations are based on machine characteristics and precedences are based on technical relationships in the BOMs of the subassemblies and the end product.

A processing time p_i, a release time r_i and a deadline d_i are associated with each of the 16 operations involved in the assembly line. The corresponding data are reported in Table 10.2.

In general, the machines' tooling and other resources for performing the operations are available at the beginning of the production plan, except for the operations j with $r_j>0$; in particular, note that in the model it is assumed that the operations which have the corresponding release date equal to zero are immediately available; otherwise, $r_j>0$ is the time period at which the resources for performing operation j become available for the assembly process. On the basis of the given weekly production plan, all the operations in the robot assembly line must be completed within the end of the second shift (6 hours/shift) of the fifth production day, i.e. the deadline of each operation is fixed to 60.

The problem is to assemble the robot while satisfying the precedence and time-window constraints and minimizing the makespan.

Table 10.2 Processing time p_j, release date r_j and deadline d_j for the assembly operations.

Operation j	p_j	r_j	d_j
1	2	0	60
2	0	2	60
3	3	0	60
4	4	3	60
5	1	0	60
6	5	0	60
7	3	0	60
8	4	2	60
9	2	7	60
10	6	0	60
11	0	0	60
12	3	3	60
13	2	9	60
14	5	19	60
15	1	5	60
16	3	0	60

Table 10.3 Time windows after the preprocessing phase.

Operation j	r_j	d_j
1	0	40
2	2	40
3	0	37
4	3	55
5	3	38
6	7	60
7	0	36
8	3	40
9	7	40
10	0	60
11	0	43
12	12	46
13	24	57
14	19	51
15	24	52
16	30	60

Starting from the data reported in Table 10.2, the application of rules (6)–(7) in the preprocessing phase has allowed the reduction of the time window of each operation to the value reported in Table 10.3. Note that starting from an overall deadline–release date gap of 910 time units (see Table 10.2), after the preprocessing phase, the time windows have been tightened to a gap of 648 time units, thus resulting in a reduction of about 28.8% over the initial constraints. According to rules (8)–(11) no reduction of the arc sets A and P has been possible. It is important to note that in the preprocessing phase, 58.3% of the input data has been updated.

Following what has been presented in the previous section, after the preprocessing phase, two dummy nodes, namely nodes 0 and 17, have been added to V in order to derive the model (12)–(22) for the robot-assembly-line problem. In particular, nodes 0 and 17 are such that $r_0 = p_0 = d_0 = 0$, $c_{0i} = 0 \; \forall i = 1, \ldots, 16$, $r_{17} = \max\{r_i + p_i,\ i = 1, \ldots, 16\} = 33$, $p_{17} = 0$, $d_{17} = \max\{d_i,\ i = 1, \ldots, 16\} = 60$, $c_{i17} = 0 \; \forall i = 1, \ldots, 16$. Now the graph of the problem consists of the sets V and A such that $|V| = 18$ and $|A| = 116$.

The optimization model of the robot-assembly-line problem can now be formulated. Let us present the model (12)–(22) related to the present case study in a suitable way for the standard MPS (Mathematical Programming System) format (IBM, 1992). The model is as follows:

$$\min\{t_{17}\} \tag{23}$$

subject to

$$t_{17} - t_v \geqslant 0 \;\; \forall v = 1, \ldots, 16, \tag{24}$$

$$t_i - t_j + (d_i + c_{ij} - r_j)x_{ij} \leqslant d_i - r_j - p_j \;\; \forall (i,j) \in A, |A| = 116, \tag{25}$$

$$t_j - t_k \geqslant b_{kj} + p_j \;\; \forall k = 1, \ldots, 16\text{: } (k,j) \in P, |P| = 19, \tag{26}$$

$$t_i - t_j + (d_i - p_i - r_j)x_{ij} \leqslant d_i - p_j - r_j \forall (i,j) \in A\text{: } c_{ij} < 0, \tag{27}$$

$$\sum_{i:(i,v) \in A} x_{iv} = 1 \;\; \forall v = 0, \ldots, 16, \tag{28}$$

$$\sum_{j:(w,j) \in A} x_{wj} = 1 \;\; \forall w = 1, \ldots, 17, \tag{29}$$

$$\sum_{i=0,\ldots 17} \; \sum_{j=1,\ldots,16} x_{ij} = 16 \;\; \forall (i,j) \in A, \tag{30}$$

$$\sum_{j=1,\ldots,16} x_{0j} - \sum_{i=1,\ldots,16} x_{i17} = 0, \tag{31}$$

$$u_i - u_j + 16x_{ij} \leqslant 15 \;\; \forall i,j = 1, \ldots, 16\text{: } (i,j) \in A, \tag{32}$$

$$r_v + p_v \leqslant t_v \leqslant d_v \;\; \forall v = 0, \ldots, 17, \tag{33}$$

$$x_{ij} \in \{0,1\}. \tag{34}$$

Note that constraints (14) are now considered as the lower and upper bounds (33) for the t_i variables. The above model consists of 236 constraints and 150 variables. Its matrix structure is reported in Figure 10.2, where columns report the t_i, u_i, and x_{ij} variables (18, 16, and 116, respectively), and rows follow the order of model (24)–(32).

The MPS file of model (23)–(34) has been generated and then executed on an IBM RISC/6000 Model 320.* By using the Optimization Subroutine Library

*Note. The MPS file of the present case study is available; interested readers can contact the author directly.

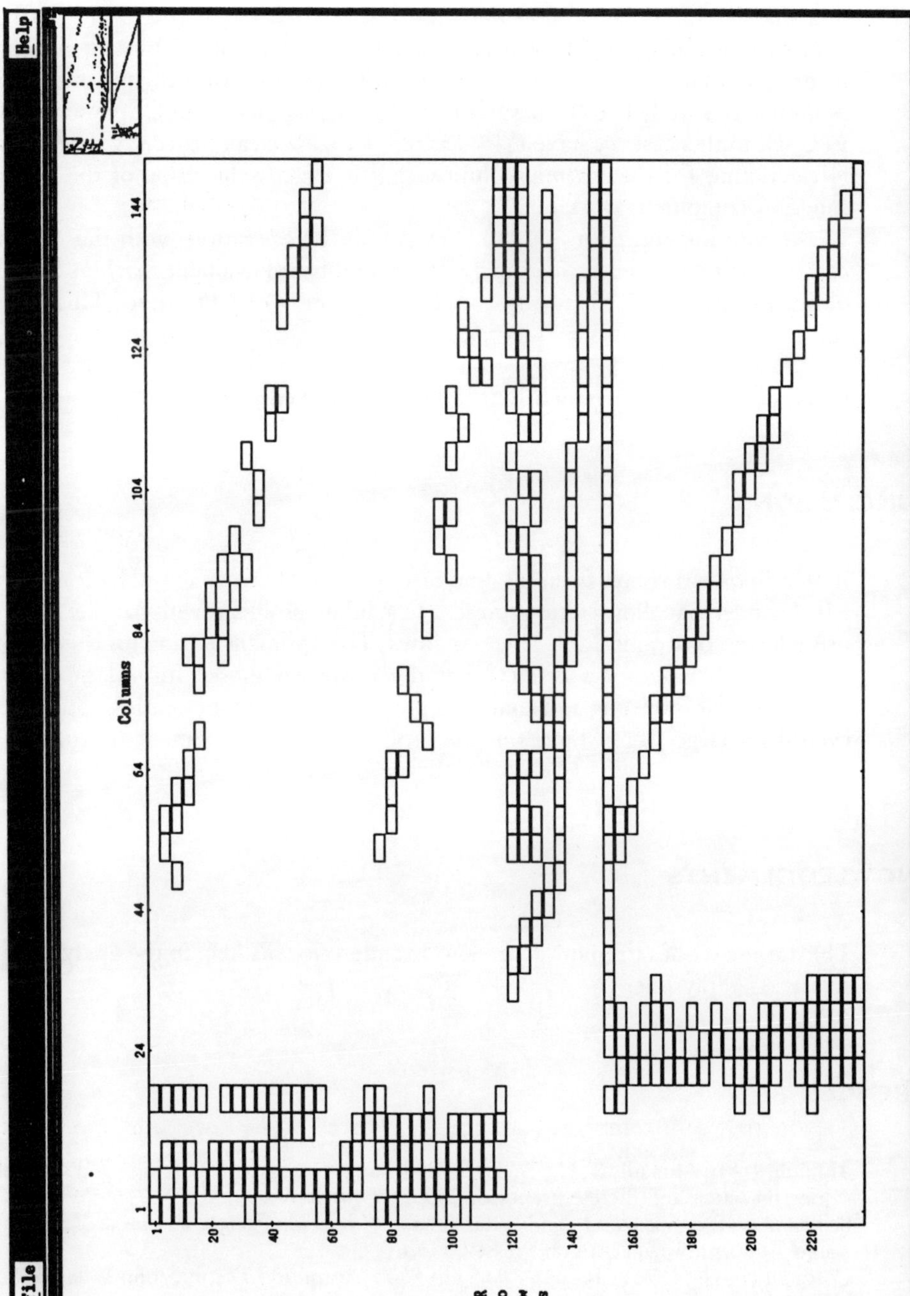

Figure 10.2

(OSL) (IBM, 1992) five different optimal solutions, corresponding to a makespan value of 33 time units, have been obtained. These solutions are related to the possibility of performing the assembling operations by using 12, 8, 5, 3, and 2 machines, respectively.

Note that looking at the objective function (23) of the problem it is possible to observe that its minimum value is given by $t_{17}=33$. In fact, by construction, node $n+1$ is added to V in such a way that $p_{n+1}=0$ and $r_{n+1}=\max\{r_i+p_i, \forall i \in V\}$; in this case we have $r_{17}=33=r_{16}+p_{16}$. We can hence say that 33 is the lower bound for the optimal solution that is exactly the value of the objective function computed by OSL.

The solution that allows us to perform all the operations with the minimum completion time and by using the minimum number of available machines has been obtained after 2277 iterations and 19.34 seconds of CPU time. This optimal solution is given by $x_{0,2}=x_{2,3}=x_{3,11}=x_{11,5}=x_{5,9}=x_{9,12}=x_{12,14}=x_{14,15}=x_{15,13}=x_{13,16}=x_{16,17}=x_{0,7}=x_{7,1}=x_{1,4}=x_{4,8}=x_{8,6}=x_{6,10}=x_{10,17}=1$. All the remaining x_{ij} variables such that $(i,j)\in A$ are fixed at 0.

5 CONCLUSIONS

In this chapter a robot-assembly-line case study has been analyzed by formulating a 0–1 model that allows us to represent scheduling problems with parallel activities, precedence constraints, and time windows. The optimal solution for this problem has been obtained. We believe that the proposed model formulation can be very useful for modeling and analyzing a wide class of manufacturing applications, even if for large-size instances of the problem only approximate approaches are possible.

ACKNOWLEDGEMENTS

The author wishes to thank Laureano Escudero for his help in the analysis of the robot assembly line.

REFERENCES

Ahmadi, J. H., Ahmadi, R. H., Dasu, S., and Tang, C. S. (1992) Batching and scheduling jobs on batch and dicrete processors, *Operations Research*, **40**, 4, 750–763.

Bellman, R., Esogbue, A. O., and Nabeshima, I. (1982) *Mathematical Aspects of Scheduling and Applications*, Pergamon Press, Oxford.

Coffman, E. G. (1976) *Computer and Job-Shop Scheduling Theory*, John Wiley & Sons, New York.

Escudero, L. F., and Sciomachen, A. (1990) *An Approximate Algorithm for the Sequential Ordering Problem with Time Windows and Precedence Constraints*, IBM Research Report RC 16280, T. J. Watson Research Center, Yorktown Heights, NY.

IBM (1992) *Optimization Subroutine Library, Reference Guide Release 2.*

Lawler, E. L., Lenstra, J. K., Rinnoy Kan, A. H. G., and Shimoys, D. B. (1985) *The Traveling Salesman Problem*, John Wiley & Sons, Chichester.

Wittrock, R. J. (1986) *Scheduling Parallel Machines with Setups*, IBM Research Report RC 11740, T. J. Watson Research Center, Yorktown Heights, NY.

Wittrock, R. J. (1988) *An Adaptable Scheduling Algorithm for Flexible Flow Lines*, *Operations Research*, **36**, 3, 445–453.

Dipartimento Scienza del'Informazione
Università degli Studi di Milano
Via Cornelico 39/41
20135 Milan
Italy

11
Task Synchronization in a Flexible System for Car Components Assembly

Alessandro Agnetis[1], **Alessandro Ciancimino**[1], **Massimo Pizzichella**[1] **and Mario Lucertini**[2]
[1]*Università di Roma 'La Sapienza', Italy*
[2]*Centro 'Vito Volterra' and II Università degli Studi di Roma, Italy*

ABSTRACT
This chapter deals with the modeling and optimization of task assignment in the assembly of particular mechanical components, in the heater of a popular model of car. While the general assembly-line balancing problem is very difficult and complex, the structure of the production process and the layout allow us to solve this problem efficiently, even in the presence of other production constraints. The problem can be formulated in terms of graph partitioning on a suitable class of graphs and solved by means of a dynamic programming approach. The computational complexity of the approach is discussed, and the detailed solution of the problem is then illustrated, comparing the quality of the solutions obtained using different objective functions and different layout structures. The solutions are evaluated in terms of the cost parameters commonly used by shop-floor management.

1 INTRODUCTION

This chapter deals with the problem of operations assignment and synchronization in a shop floor where a set of *workstations* carry out assembly operations. The structure of the production system is a *pipeline* (also called *flow line* or *flow shop*) in which components are moved by a one-way transportation system (usually, a conveyor), and workstations are totally ordered (pipelined), so that parts and subassemblies can only move from workstation M_j to M_{j+1} (see Figure 11.1). Each workstation is able, in principle, to perform any of the required operations (it is a *general-purpose machine*).

Optimization in Industry 2. Edited by T. A. Ciriani and R. C. Leachman

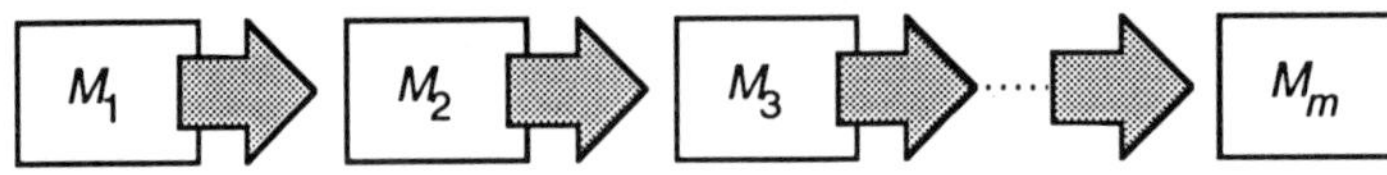

Figure 11.1 Scheme of a pipeline production system.

The advantages of pipelines include low cost, high robustness, and ease of management and supervision; they are inherently deadlock-free, and match particularly well the structure of some classes of process plans, as explained below. The drawbacks may be poor fault tolerance and routing flexibility (Askin and Standridge, 1993). In production planning problems, the process plan of the product can be described by a directed acyclic graph in which nodes correspond to operations and arcs express precedence constraints between the corresponding operations. This graph will be referred to as *operation graph G*. A pipeline organization is quite common whenever certain subsets of operations are totally ordered, i.e. *sequences* of operations can be recognized in the operation graph. Such a situation indeed occurs in a large class of assembly processes, as in the case analyzed here. A typical example of this is the assembly of several 'side components' on a main component (e.g. in electronic assembly, the chips are inserted on a circuit board); in this case the structure of the assembly graph is often a *comb*, and this allows us to design particular solution procedures (Agnetis *et al.*, 1991). Here we are concerned with a situation in which the operation graph consists of a *forest* of two chains plus some additional arcs (Section 2); the same methodology can, however, be generalized to an arbitrary number of chains, with an appreciable increase in the computational complexity. From an engineering viewpoint, the two chains correspond to operations on a *main path* and other *secondary operations*. Other constraints exist on the operation assignments to the m workstations, as described in Section 2.1.

The problem consists of assigning the operations to the m machines. In this chapter the operation assignment problem is solved by means of a graph-partitioning model. In fact, an assignment defines an m-partition of the nodes of the operation graph.

The goal is to maximize productivity, or minimize overall production costs. These goals are often expressed in terms of balancing, as far as possible, the workload among the workstations, since a workload-balanced solution usually attains the highest utilization and minimizes the bottlenecks. Moreover, the *number* of workstations affects the overall costs (Section 4). On the other hand, some layouts allow for a certain degree of *parallelism*: some workstations can be duplicated and perform in parallel the same operations on two different units.

The chapter is organized as follows: in Section 2, a more detailed description of the problem is given, discussing the constraints holding on feasible partitions, and the objective functions introduced. In Section 3, the solution algorithms are illustrated. In Section 4, a detailed analysis of a case study is carried out: the application concerns the assembly of a heater for small cars. Two different layout situations are considered, the optimization model is applied in both cases, and the solutions are analyzed.

2 THE TASK-ASSIGNMENT PROBLEM

The problem concerns the cyclical production of a single product type. The processing of each *unit* consists of the execution of n operations $\{O_1, O_2, \ldots, O_n\}$, which must be assigned to m *workstations* $\{M_1, M_2, \ldots, M_m\}$. A precedence relation among operations is expressed by an *operation graph* $G=(N,A)$, in which each node corresponds to an operation and arc (i,j) indicates that O_j can only start after the completion of O_i. Node i is weighted by the *length* τ_i of operation O_i. The overall flow-management problem in assembly systems consists of assigning the operations required by each unit to the machines (*routing problem*), and scheduling the subassemblies (possibly belonging to different units) on the machines (*scheduling problem*). Since our system cyclically produces identical units, only the former problem is relevant, and it will be addressed here.

The problem consists of finding an m-partition $\Pi \equiv \{\pi_1, \pi_2, \ldots, \pi_m\}$ of the nodes of G, where π_j consists of the operations assigned to machine M_j. For an assignment of operations to machines to be feasible, some conditions must hold on Π: since the system is a pipeline, subassemblies visit the machines at most once; therefore if an arc (i,h) exists in G, we cannot assign i to M_j and h to M_k, if $k<j$. In other words, given a partition Π, we associate with it a graph $G_\Pi=(N_\Pi, A_\Pi)$ as follows: $|N_\Pi|=m$; $(j,k)\in A_\Pi$ if there is at least one arc $(i,h)\in A$ such that $i\in\pi_j$ and $h\in\pi_k$. For a partition Π to be feasible, the graph G_Π must be acyclic. (Note that any topological numbering of the nodes of G_Π yields a feasible operations-to-machines assignment.)

The best-known model for routing in assembly lines is probably the so-called *Assembly-line balancing* problem, which consists in partitioning the nodes of the operation graph into a minimum number of classes such that the sum of the operation times in each class does not exceed a given bound (*cycle time*). For a survey of this problem, see, for instance, Ghosh and Gagnon (1989). Assembly-line balancing is in general a hard problem; in fact, it is a generalization of *bin packing* (obtained if $A \equiv \{\varnothing\}$). However, in some practical cases, the structure of the process plan is such that the problem can be solved in polynomial time. In many process plans, there is a 'main path' of operations, consisting of operations that must be sequentially performed on some main component of the product; these often involve the same subassembly. Besides these operations, there are other *secondary* operations that must be performed; these are not constrained to a particular position on the main chain, but for each of them a range of possible positions is specified. For example, in Figure 11.2 the secondary operation i cannot be scheduled before operation h and later than operation k. Also secondary operations are totally

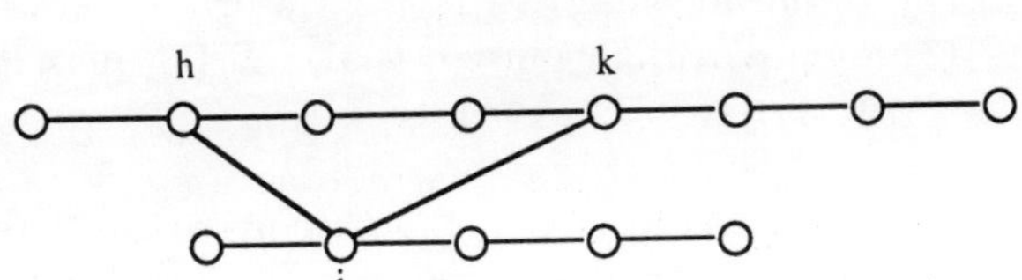

Figure 11.2 Constraints on secondary operations.

ordered. Hence, the operation graph consists of two chains plus some additional arcs. In the following, we explain further conditions that may restrict the set of feasible partitions of graph G, and introduce the objective functions employed.

2.1 Restrictions to feasible partitions

We next consider two types of constraints for the partition Π, namely *incompatibilities* between operations and *mandatory assignments* of operations to workstations.

Two operations are *incompatible* if they cannot be assigned to the same workstation. Typically, this occurs because the size of the tooling and the equipment required by the operations exceeds the physical space assigned to the workstation. In this case, the two operations must be assigned to different workstations. Note that two incompatible operations may not be adjacent in the operation graph. Actually, one can observe that such an incompatibility relation should in general be defined for k-tuples of operations: so, for instance, three operations might be pairwise compatible, but not compatible as a group of three. For the sake of simplicity, we disregard such situations, but this extension is not a difficult concept.

For reasons still related to the physical space taken up by the equipment, certain tools and/or fixtures might be assigned *a priori* to a given workstation, and it might be unprofitable or impossible to move them. In this case, the operations requiring that piece of equipment *must be assigned* to that workstation (*mandatory assignments*). Note that if two operations on the same chain must be assigned to a workstation, the same must be true for all the operations in between.

2.2 The objective functions

The objective of the task assignment is to balance, as far as possible, the workload among the machines. This can be expressed by means of several different objective functions. In fact, it is sometimes more important to distribute the workload evenly among the workstations, rather than having a shorter cycle time with a greater unbalancing between workstations and per-unit labor cost (Section 4).

In the following, let $W(\pi_j)$ be the total weight of the set π_j of operations assigned to machine M_j, and μ be the average cluster weight, i.e. $\mu = (\Sigma_j W(\pi_j)/m)$. In this chapter we consider three objective functions:

(1) L_1 norm: $\min z = f_1(\Pi) = \Sigma_j |W(\pi_j) - \mu|$,
(2) L_2 norm: $\min z = f_2(\Pi) = \Sigma_j (W(\pi_j) - \mu)^2$,
(3) L–U partition: $\min z = f_3(\Pi) = \Sigma_j [c_U \max \{0, W(\pi_j) - U\} + c_L \max \{0, L - W(\pi_j)\}]$.

The first two objective functions correspond to the usual goal of balancing the workloads among the machines. The solutions produced by these objective functions are nonetheless different in terms of some evaluation indices commonly used in practice (see Section 4).

The third objective function realizes a slightly different criterion: we search for a task assignment such that the workload of each machine is greater than L and smaller than U. All the solutions satisfying this requirement are equally optimal. This corresponds to specifying a maximum allowed cycle time (U) and a minimum amount of work per workstation (L). If such a solution does not exist, the amount by which the requirements are violated is weighted by the coefficients c_U and c_L. Usually, not exceeding U is more important than having a minimum work amount L, i.e. it can be $c_U > c_L$. Clearly, the four parameters L, U, c_L, c_U can be appropriately tuned.

2.3 Problem statement

In conclusion, the task assignment problem can be stated as follows:

> Given a pipeline system consisting of m machines, given a part type and its corresponding operation graph, given a set of incompatibilities between operations and mandatory assignment constraints, find a feasible assignment of operations to machines Π such that objective function $f(\Pi)$ is minimized.

Here $f(\Pi)$ is (1), (2), or (3).

3 DYNAMIC PROGRAMMING APPLIED TO A TWO-CHAIN PARTITIONING PROBLEM

3.1 Introduction

Here we briefly review some basic ideas of dynamic programming; one of the several books on the subject is the one by Denardo (1982). Dynamic programming is a technique that can be helpfully applied to some problems in which a partial enumeration of the feasible decisions, namely the values of a set of variables, is required. Such a set of variables is referred to as the system state s, and is an element of the state-space set S.

The following information is associated with each state s:

- the state value $f(s)$; i.e. the optimal objective function value reached by a partial sequence of decisions that lead to the state s;
- information needed in order to backtrack an optimal partial sequence;
- a set $D(s)$ of decisions, among which the next decision d must be chosen, leading to the state $t(s,d)$ with a cost $r(s,d)$.

The problem is hence to define a recursive equation (*functional equation*) that, starting from a state s with the objective-function value $f(s)$, recursively allows us to choose the decision d leading towards that particular next state x such that the new objective-function value $f(x)$ is still optimal. In other words:

$$f(x) = \min\{r(s,d) + f(s) | s \in S\}$$

where $r(s,d)$ is the cost associated with decision d, chosen at the state s. At each subsequent iteration, for each state s, among all the decisions leading to such a state, only the optimal will be stored, discarding the others.

In order to apply dynamic programming, the *optimality principle* must hold, namely that if $d_1,d_2,\ldots,d_n$ is a sequence of decisions leading to the optimal solution then the last k $(1\leqslant k\leqslant n)$ decisions must be optimal. In other words a sequence of optimal decisions must be optimal.

3.2 Partition of two parallel chains

Consider now an operation graph G consisting of two parallel chains, denoted as A and B (see Figure 11.3). The problem is to find a feasible q-partition (i.e. a partition in q clusters) of G. Let n_A and n_B be the numbers of nodes respectively of chains A and B. Let W_A and W_B be the node-weight arrays respectively of chains A and B. Let $W(\pi_j)$ be the weight of the jth cluster (including nodes from both chains A and B). Let S_{k_A} (S_{k_B}) be the subchain of chain A (B) consisting of the first k_A (k_B) nodes.

The average cluster weight is:

$$\left(\sum_{i=1}^{n_A} W_A(i)+\sum_{i=1}^{n_B} W_B(i)\right)\Big/q$$

Let (k_A,k_B,q) be the state s, where:

- k_A (k_B) is the rightmost node of chain A (B), belonging to the subchain S_{k_A} (S_{k_B}).
- q represents the number of clusters in which the subgraph consisting of subchains S_{k_A} and S_{k_B} has to be partitioned.

Also, let $\langle u_A,u_B,k_A,k_B\rangle$ denote a cluster consisting of subchain (u_A,k_A) of chain A and (u_B,k_B) of chain B.

3.3 The algorithm

In the following, we describe an algorithm for the optimal feasible p-partition of a graph consisting of two chains. The objective function can be any of those introduced in Section 2.2.

In the statement of the algorithm, $c(u_A,u_B,k_A,k_B)$ is the cost of cluster $\langle u_A,u_B,k_A,k_B\rangle$; for ease of notation, in the following, whenever $k_A<u_A$ $(k_B<u_B)$,

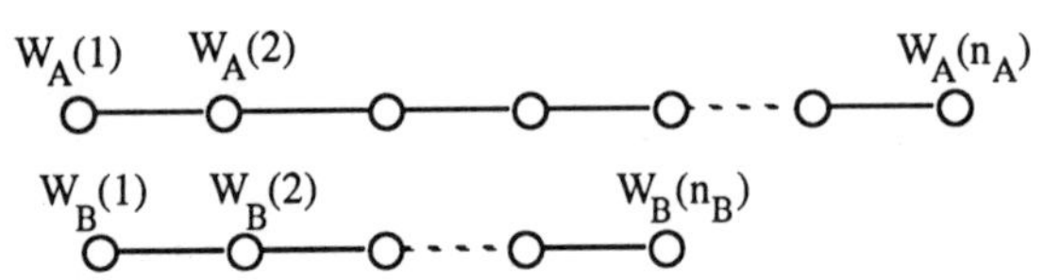

Figure 11.3 Graph G consisting of two parallel chains.

we mean that the cluster does not contain nodes of chain A (B). Supposing that $\langle u_A, u_B, k_A, k_B\rangle$ is the qth cluster of the partition, and recalling Section 2.2, $c(u_A, u_B, k_A, k_B)$ is given by

- $c_A \max\{0, L - W(\pi_q)\} + c_B \max\{0, W(\pi_q) - U\}$ for the $(L-U)$ partition;
- $|W(\pi_q) - \mu|$ for the L_1-norm partition;
- $(W(\pi_q) - \mu)^2$ for the L_2-norm partition.

Let $G_{k_A k_B}$ be the graph consisting of subchains S_{k_A} and S_{k_B}. In what follows, $f(k_A, k_B, q)$ is the optimal value of a q-partition of $G_{k_A k_B}$. If $k_A = 0$ ($k_B = 0$) $G_{k_A k_B}$ only contains subchain S_{k_B} (S_{k_A}).

Algorithm FIND_PARTITION
Begin
  for $k_A := 1$ to n_A **do**
    for $k_B := 1$ to n_B **do**
      $f(k_A, k_B, 1) = c(1, 1, k_A, k_B)$;
  for $q := 2$ to p **do**
    for $k_A := 1$ to n_A **do**
      for $k_B := 1$ to n_B **do**
  if $k_A + k_B \geqslant q$ **then**
    $f(k_A, k_B, q) := \mathbf{min}\{c(u_A, u_B, k_A, k_B) + f(u_A - 1, u_B - 1, q - 1)\}$
        with respect to $\{u_A \in [1, k_A + 1], u_B \in [1, k_B + 1]\}$;
  if $k_A + k_B < q$ **then**
    $f(k_A, k_B, q) := \infty$;

end FIND_PARTITION.

For each state, besides the optimal value $f(k_A, k_B, q)$, the corresponding partition is stored, in order to be able to backtrack the optimal solution.

The complexity of FIND_PARTITION

In order to evaluate the complexity of the algorithm illustrated in the previous section, we note that the computation of $f(k_A, k_B, q)$ is done in $O(n_A n_B \log n)$, since u_A and u_B may assume $O(n_A)$ and $O(n_B)$ different values respectively. Such computation is performed $O(n_A n_B p)$ times, as is apparent from the nested structure of the algorithm. As a consequence the algorithm has complexity $O(n_A^2 n_B^2 p)$.

3.4 The constraints

If additional constraints exist, these can be easily incorporated in the algorithm of Section 3.2. In fact, if some cluster $\langle u_A, u_B, k_A, k_B\rangle$ contains incompatible node pairs, then we only have to set $c(u_A, u_B, k_A, k_B)$ to infinity. Similarly, we can model

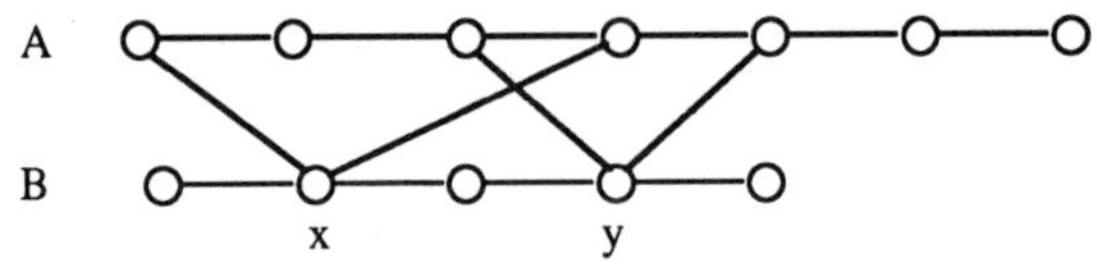

Figure 11.4 Precedence constraints across the two chains.

mandatory assignment constraints by an infinite value of $c(u_A, u_B, k_A, k_B)$, if an operation $i \in \langle u_A, u_B, k_A, k_B \rangle$ must be assigned to a machine M_j, and such a cluster cannot correspond to machine M_j.

Most often, as in the application described in Section 4, precedence relations exist also *across* the chains A and B (Figure 11.4). We may follow exactly the same scheme of the algorithm FIND_PARTITION, provided that care is now taken with the feasibility check. In fact, when evaluating the cost of a solution, we must check whether or not the graph G_Π obtained by appending the cluster $\langle u_A, u_B, k_A, k_B \rangle$ to the graph G_{u_A-1, u_B-1} is acyclic. This would be, for general graphs, rather costly in terms of computational efficiency. However, due to the particular structure of our graph, it can be easily shown that we can limit ourselves to checking whether or not the cluster $\langle u_A, u_B, k_A, k_B \rangle$ has no outgoing arcs: if there are outgoing arcs, the cost $c(u_A, u_B, k_A, k_B)$ can be set to $+\infty$. Figure 11.4 shows an example of precedence relations of nodes x and y.

4 THE APPLICATION

In this section, we present the application of the models illustrated in the previous sections to the assembly of a heating device for medium-powered cars. In Section 4.1 the product and the process are briefly described; in Section 4.2 two different layout solutions are presented and the preprocessing of the model is illustrated; in Section 4.3 the numerical results are shown. Several details of the numerical experience can be found in Pizzichella (1989).

4.1 The product, the process, and the initial solution

The product consists of four major components:

- the *box*, composed of two black plastic *shells*, containing other elements (baffles for mixing cold and warm air, command cables, electric wires, an electric motor-fan, a radiator);
- the *console* (visible from the interior of the car), on which two mouths, their wickets and the control panel are set up;
- the *control panel* with its leverages and knobs;
- the *radiator*; made of aluminum, consisting of some elliptic section tubes for optimal thermal exchange, and some other protective and insulation elements.

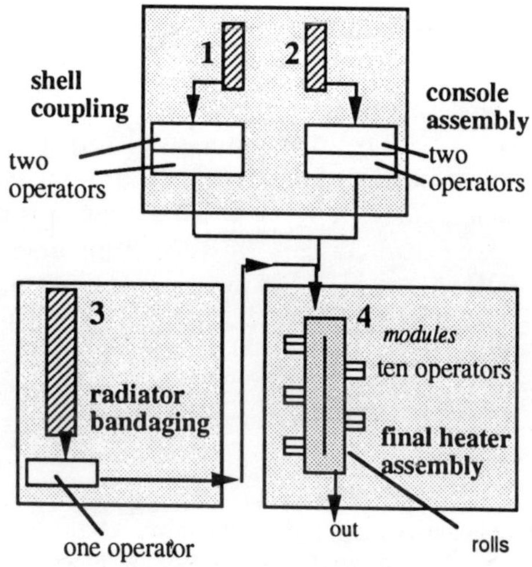

Figure 11.5 The actual plant layout.

In our case study, a single product type is produced in a day of production. Figure 11.5 shows the layout of the system before optimization took place. This layout was shaped according to the process structure: the shells, the console and the radiator are prepared in distinct areas of the plant and put together in the final assembly process. Drawbacks of this solution are that the workstations are far away from each other, there are frequent conflicts in the transportation network (consisting of several baskets moved by trolleys, very often in motion between the workstations), and above all the segmentation of the process eliminates many opportunities for a better operation assignment. Moreover, since it is not possible to achieve perfect workload balancing between different lines, a considerable work-in-process arises.

Most of the assembly tasks are manually performed by low-skill workers using low-cost tools, such as pneumatic screwdrivers and riveters; the only exception concerns shell coupling. The shells and the console are obtained by two different moulding presses (numbered 1 and 2 in Figure 11.5): after some finishing operations, the shells are coupled while they are still warm from the pressing, using a Bronson vibration welder. The whole shell coupling subprocess is carried out by two sequential workstations. In parallel, the console comes out from the other moulding press, and another two sequential workstations perform the mounting on the console of some levers, pulleys, wickets, linings, and an object holder; moreover, plugs and wickets are set up, and other preparation tasks are performed by a specific pneumatic tool. Radiators arrive already assembled at the bandaging workstation (numbered 3), whereas the control panel is assembled in a different plant. The final phase of the assembly process is performed in parallel on five identical *modules* (numbered 4), each consisting of two sequential operators. The operation assignment is the same for all modules. The final phase consists in: assembling the radiator and the console, assembling the console and the control panel with the front panel, electric wiring, arranging and regulating all the command cables, and some other secondary tasks. The modules are dislocated along a tension roller on which the product is transported. At the end of every assembly phase its quality is controlled visually.

The actual assembly process employs 15 workers (2 in shell coupling, 2 in console assembly, 1 in radiator bandaging, 10 in final assembly). The annual demand is 137 500 units, corresponding to an average productivity of 625 heating devices per day. Daily demand can, however, fluctuate between 550 and 800 units.

In the following tables, the process plan and the original operation assignment are described. For each task we indicate the task number, a brief description of

Table 11.1 Shell coupling—operator 1.

Task no.	Task	Length of task (s)	Tool required
1	Shells collecting and finishing	14	Knife
2	Shells positioning on Bronson welder	10	Bronson
3	Lids positioning and start	8	Bronson
4	Collecting and table positioning on table	6	Bronson
5	Baffles and motor-fan positioning and fixing with springs	34	
6	Parts supplying	6	
Workload (per unit):		78	

Table 11.2 Shell coupling—operator 2.

Task no.	Task	Length of task (s)	Tool required
7	Shells collecting	5	
8	Five self-tapping screws are inserted	18	Pneumatic screwdriver
9	Four mouths setting up	12	
10	Bracket fixing with two rivets	17	Pneumatic riveter
11	Deflectors positioning	10	
12	Packing	6	
13	Parts supplying	6	
Workload (per unit):		74	

Table 11.3 Console assembly—operator 1.

Task no.	Task	Length of task (s)
14	Console collecting and quality control	7
15	Lever collecting and assembly	6
16	Pulley collecting and assembly	6
17	First wicket collecting and assembly	6
18	Second wicket collecting and assembly	6
19	Lining collecting and assembly	5
20	Object-holder collecting and assembly	6
21	Console packing and storing	18
Workload (per unit):		60

Table 11.4 Console assembly—operator 2.

Task no.	Task	Length of task (s)	Tool required
22	Console collecting and unpacking	7	
23	Visual quality control and leverages control	7	
24	Right wicket second assembly	6	
25	Left wicket second assembly	6	
26	Wickets functional control	5	
27	Plugs inserting preparation	4	Pneumatic tool
28	Plugs inserting	5	
29	Cleaning operations	5	
30	Console packing and storing	10	
31	Parts supplying	4	
Workload (per unit):		60	

Table 11.5 Radiator protective bandaging.

Task no.	Task	Length of task (s)
32	Radiator collecting and positioning	4
33	First band collecting and preparation	6
34	First band sticking on radiator	5
35	Second band collecting and preparation	6
36	Second band sticking on radiator	5
37	Radiator packing and storing	3
Workload (per unit):		29

Table 11.6 Module assembly—operator 1 (there are five in parallel).

Task no.	Task	Length of task (s)	Tool required
38	Box collecting	5	
39	Radiator collecting and positioning	7	
40	Two self-tapping screws inserting on radiator	10	Pneumatic screwdriver
41	Box assembling with twelve springs	66	
42	Control-panel collecting and quality control	7	
43	Leverages setting up on control-panel bowden	10	Pneumatic screwdriver
44	Control panel and box assembling	17	Pneumatic screwdriver
45	Bowden assembly and wicket regulation	27	Pneumatic screwdriver
46	Bowden assembly and admission-wicket regulation	24	Pneumatic screwdriver
47	Bowden assembly and tap regulation	24	Pneumatic screwdriver
48	Final control-panel functionality control	14	
49	Faston collecting, assembly and wiring	7	
50	90° rotation	3	
51	Stamping	7	
52	Superior lining sticking	7	
53	Slotted-cable setting up	7	
54	Cable inserting in the slotted cable	5	
55	Water-pipe inserting	10	
56	Lining sticking	12	
57	Unloading	3	
Workload (per unit):		272	

Table 11.7 Module assembly—operator 2 (there are five in parallel).

Task no.	Task	Length of task (s)	Tool required
58	Heating-device collecting	3	
59	Faston-motor coupling	3	
60	Console collecting and unpacking	10	
61	Console visual control	14	
62	Console and heating-device assembling with four screws	24	Pneumatic screwdriver
63	Leverages setting up on the right bracket	5	
64	Right-bracket setting up on heating device	10	Pneumatic screwdriver
65	Two screws setting up on one side of the console	14	Pneumatic screwdriver
66	Lighter assembling on front panel	8	
67	Faston and front-panel assembly	7	
68	Brightness control	14	
69	Front panel and heating-device assembly with two screws	15	Pneumatic screwdriver
70	Four knobs and control-panel assembly	14	Thermo-fan
71	Wiring adjustment	7	
72	Motor-speed leaking control and adjustments	56	
73	Radiator control	24	
74	Central-bracket fixing with two rivets	21	Pneumatic riveter
75	Stamping	8	
76	Lining sticking	10	
Workload (per unit):		267	

the operation, its length (in seconds), and the tool required. We refer to these task numbers throughout the paper.

In the actual solution, the bottleneck turns out to be the first shell-coupling operator, yielding a cycle time of 78. Therefore, if daily production is organized into two 7.5 hour shifts, this corresponds to a daily production capacity of $(3600 \times 15/78) = 692$ units.

4.2 Application of the model

The original layout was subdivided into areas, corresponding to three major components (shells, console, radiator) and to the final-assembly phase. We have already observed that this restricted the possible operations assignments; therefore it seems profitable to consider alternative layouts, possibly allowing a higher degree of flexibility in the assignment of operations to workstations. In this case study we analyzed two different solutions, in the following referred to as *general-lines solution* and *dedicated-lines solution*.

The general line is a flow line; all the assembly tasks are executed in sequence on the same line; possibly, the whole line can be duplicated to obtain a better balancing and a higher cycle time on a single line, if this is convenient from the overall-cost viewpoint. The general-lines solution allows us to eliminate most of the packing and unpacking tasks, and reduce the work-in-process.

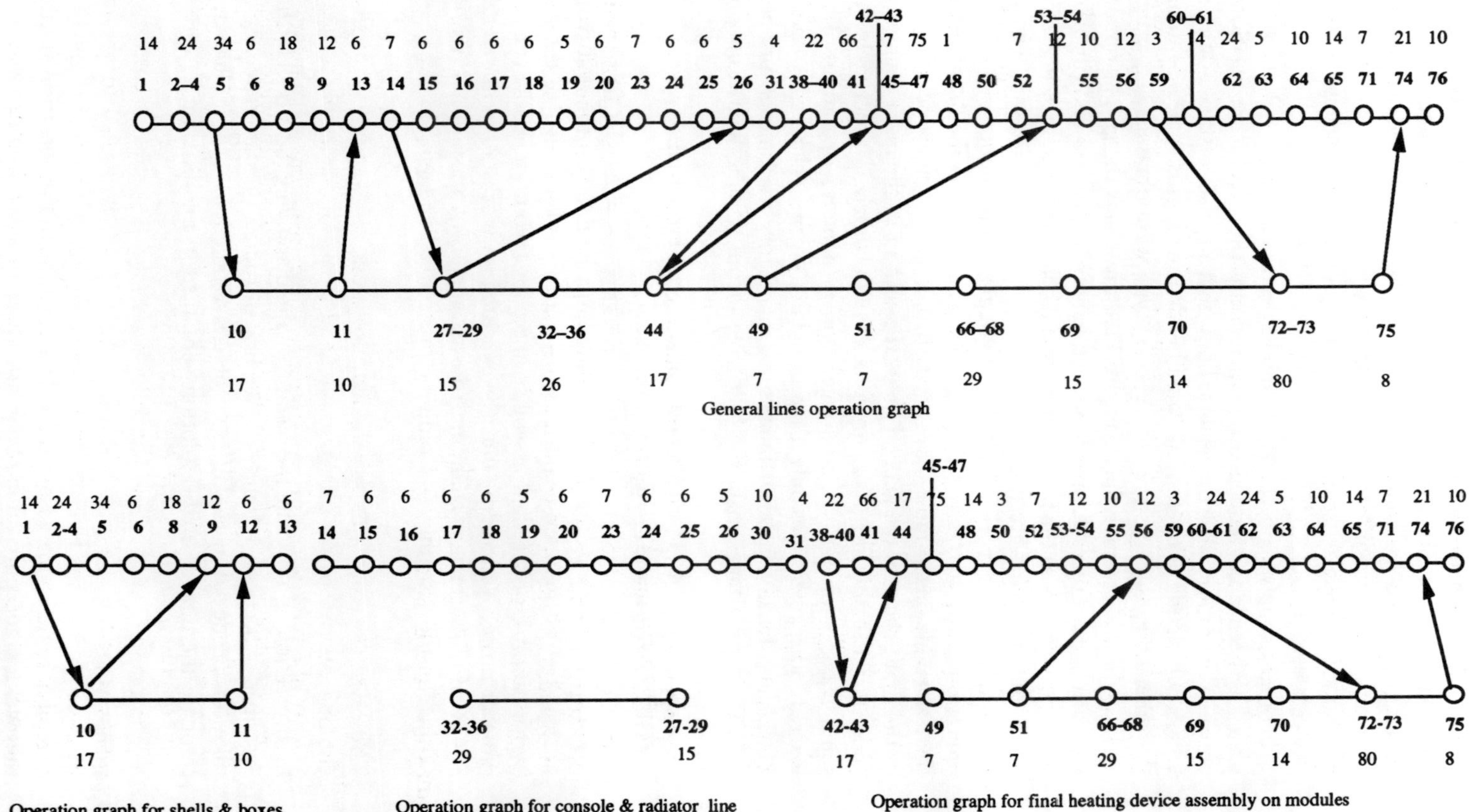
42–43
53–54
60–61
14 24 34 6 18 12 6 7 6 6 6 6 5 6 7 6 6 5 4 22 66 17 75 1 7 12 10 12 3 14 24 5 10 14 7 21 10
1 2–4 5 6 8 9 13 14 15 16 17 18 19 20 23 24 25 26 31 38–40 41 45–47 48 50 52 55 56 59 62 63 64 65 71 74 76
10 11 27–29 32–36 44 49 51 66–68 69 70 72–73 75
17 10 15 26 17 7 7 29 15 14 80 8
General lines operation graph
45-47
14 24 34 6 18 12 6 6 7 6 6 6 6 5 6 7 6 6 5 10 4 22 66 17 75 14 3 7 12 10 12 3 24 24 5 10 14 7 21 10
1 2-4 5 6 8 9 12 13 14 15 16 17 18 19 20 23 24 25 26 30 31 38-40 41 44 48 50 52 53-54 55 56 59 60-61 62 63 64 65 71 74 76
10 11
17 10
32-36 27-29
29 15
42-43 49 51 66-68 69 70 72-73 75
17 7 7 29 15 14 80 8
Operation graph for shells & boxes
Operation graph for console & radiator line
Operation graph for final heating device assembly on modules

Figure 11.6

Dedicated lines are halfway between the original layout and the general lines. They partially embody some physical and assembly constraints, but the flexibility of the overall task assignment is limited. In this work hypothesis, we arrange the production on three different lines, respectively taking care of: (1) shells and boxes assembly; (2) console assembly and radiator protective bandaging; (3) final heating-device assembly on modules.

In order to apply the optimization model described in Section 3, the actual process plan has been analyzed, in order to identify the precedence relations existing between operations. For the general lines, this analysis yielded a unique graph in which two chains of operations could be identified, plus some additional arcs; in the other case, three graphs of the same type were obtained, separately for the three dedicated assembly lines. Figure 11.6 illustrates the graphs obtained in the two scenarios. For each operation, the length is specified and (in boldface) the correspondence with the former operation labeling is given. Precedence constraints across the two chains are also indicated. One can observe that some groups of tasks requiring the same workstation were aggregated into unique tasks. For instance, operations 2 through 4 (positioning of shells and lids on the Bronson welder and their assembly on the table) have been 'shrunk' into a single operation: the second of the main chain in the general lines, the second of the main chain in the shell-and-boxes assembly line. As a consequence of the different layout, in both scenarios some transfer operations no longer apply (namely, operations 37, 57, 58) whereas other operations only disappeared in the general-lines case (operations 12, 21, 22, 30). Finally, incompatibilities and mandatory assignments must be specified.

In the general-lines case, the following incompatibility constraints hold:

tasks **13** and **14**, **31** and **38–40** on the primary chain;
tasks **27–29** and **32–36** on the secondary chain;

For quality reasons the shells welding must be done on the first workstation, so task **2–4** of the main chain must belong to workstation 1 of the general line (workstation 1 of the shell assembly line); for layout and conveyor reasons the radiator protective banding must be executed on workstation number 4, so task **32–36** of the secondary chain must be assigned to workstation 4 (workstation 1 of the console-assembly and radiator protective-bandaging line).

4.3 Results

In this section we illustrate the results obtained applying the model to the product described in Sections 4.1 and 4.2. We first introduce the parameters by which the solutions produced by the different objective functions were evaluated; then, we compare the quality of the solutions obtained in the different cases.

Evaluation parameters

Several parameters are associated with a partition of the operation graph. In what follows, the *costs* are all expressed in terms of *time*: in other words, costs are regarded as proportional to the overall time spent by the workers in the production process. For a given partition, the *cycle time* T_c is the time assigned to the bottleneck workstation; its reciprocal $1/T_c$ is the *throughput*. Let Q indicate the *required* number of units produced in one shift, and D the length of one shift (here we suppose $D = 7.5$ hours). We introduce the following:

- *ideal per-unit labor cost* (ULC_{id}) is the minimum theoretical time each unit stays in the system; it is given by the sum of all the operation times:
$$\text{ULC}_{\text{id}} = \sum \tau_i;$$
- *desired cycle time* (T_{cd}): the cycle time necessary to produce Q units, i.e.
$$T_{\text{cd}} = D/Q$$
- *actual production* (Q_{p}): the number of units actually produced in one shift:
$$Q_{\text{p}} = D/T_{\text{c}};$$
- *total available time* (T_{a}): in a system with m workstations, it is given by
$$T_{\text{a}} = mD;$$
- *per-unit labor cost* (ULC): labor time devoted to each single unit produced:
$$\text{ULC} = T_{\text{a}}/Q_{\text{p}} = T_c m;$$
- *line efficiency* (E) is an index of the quality of the solution:
$$E = \text{ULC}_{\text{id}}/\text{ULC};$$
- *efficiency of workstation* $M_j(E_j)$ measures the relative unbalancing among workstations: E_j is defined as
$$E_j = w(\pi_j)/T_{\text{c}}.$$

Notice that these parameters highlight different aspects of the solution. For instance, clearly minimizing T_c also minimizes ULC; however, partitions with a slightly higher value of ULC, but smaller values for the E_j's may be preferable. Moreover, notice that the total number of workstations m, when considered as a decision variable, plays a major role. Therefore, we employed the three objective functions introduced in Section 2.2 and compared the solutions obtained in terms of these evaluation indices.

The partitions

While norms L_1 and L_2 do not require any choice for the parameters of the optimization, in order to solve the problem with the L–U objective function, the constants L, U, c_U and c_L had to be tuned. One obviously desires a partition with L and U as close as possible: if L is much smaller than U, it is easier to find a feasible partition (from the viewpoint of L and U) but not a 'good' one (i.e. being highly unbalanced is possible). In the experiments done, U was set to the desired cycle time T_{cd} (therefore usually the optimal solution had a nonzero cost), and L to $0.7U$, $0.8U$ and $0.95U$; c_L was set to 1, c_U to 10.

General lines

In the following we refer to the operation graph of Figure 11.6. In this case, adding up all the operation times, we get $\text{ULC}_{id} = 775$ seconds. Let us consider the average production requirements of $Q = 625$ units/shift. If we considered a *single* general line, desired cycle time would be $(3600 \times 7.5)/625 = 43.2$ seconds. This is clearly too small to get feasible solutions, and therefore *two* identical lines (with 10 workstations each) were considered. In order to get the same productivity, desired cycle time on each line is therefore set to $T_{cd} = 86$ seconds.

L–U partitions	
$L = 0.7U$	
Objective-function value	0
Cycle time	86
Labor cost per unit (seconds)	860
Sub-line efficiency	90%
$L = 0.8U$	
Objective-function value	0
Cycle time	85
Labor cost per unit (seconds)	850
Sub-line efficiency	91%
$L = 0.95U$	
Objective-function value	41
Cycle time	85
Labor cost per unit (seconds)	850
Sub-line efficiency	91%
L_1-norm partition	
Cycle time	86
Labor cost per unit (seconds)	860
Sub-line efficiency	90%
L_2-norm partition	
Cycle time	85
Labor cost per unit (seconds)	850
Sub-line efficiency	91%

In this case, the best partitions are obtained using the L–U objective function with $L = 0.8U$ and $0.95U$; and using the L_2 norm (the latter two partitions are the same). These partitions show the same value for what concerns labor cost per unit, cycle time and minimum workstation efficiency (69/85). The L_2 norm is preferred because of the smoother distribution of workload over the workstations (measured, for instance, by the L_2 norm itself). The partition is shown in Figure 11.7.

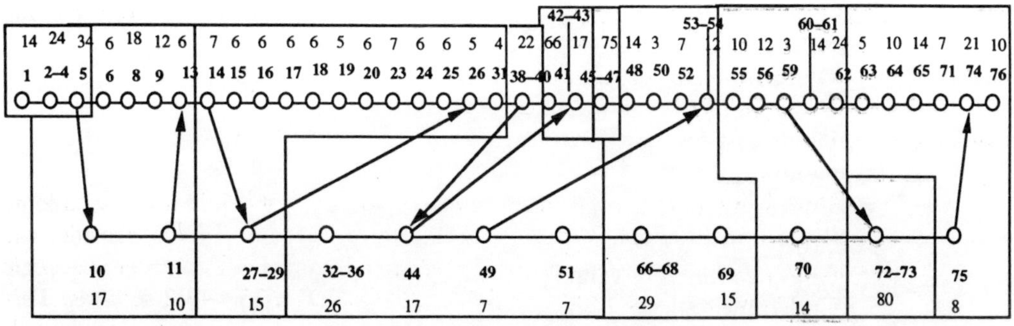

Figure 11.7

Dedicated lines

In the case of the dedicated lines, some decisions must be made prior to the application of the model. In fact, we wish to avoid, as far as possible, accumulation of material between the lines. To this aim, the cycle times of the three lines must not be too different. The problem here is to balance the final heating assembly line: in the original architecture, five modules, with two operators each, perform in parallel the same operations on five different units. If we increase the number of operators to *three* per module, the productivity increases sharply: the cycle time on each module is now smaller, and the increase in the number of units that can be produced is 'amplified' by the factor 5. In fact, with two operators per module one can get 501 units/day; with three operators 754 units/day. Since we are interested in a productivity of 625 units/day, we consider a situation in between, namely three modules with three operators, and two modules with two operators. In this way, since the total work content of the final assembly line is 537 seconds, a lower bound on the cycle time is 537/13 = 41. The total work contents of the shell-and-boxes assembly line and console assembly line are 147 and 124 seconds respectively; this suggests the assignment of four and three operators respectively to the two lines, in order to get the lower bounds of 36 and 41 seconds for the cycle time of the sub-line. In this way, the same total number of operators (20) as in the general-lines case is involved, and for each sub-line there is the theoretical possibility of attaining the desired cycle time (43 seconds).

The desired cycle time is 43 seconds; the value of ULC_{id} is different for the different sub-lines. The optimization models have been applied separately to each sub-line, yielding the following results.

Shells-and-boxes assembly line

Labor ideal cost per unit (seconds)	147
Number of operators	4
Total working time (seconds)	108000

L–U partitions

$L = 0.7U$

Objective-function value	0
Cycle time	41
Labor cost per unit (seconds)	164
Sub-line efficiency	90%
Units/shift	658

$L = 0.8U$

Objective-function value	0
Cycle time	41
Labor cost per unit (seconds)	164
Sub-line efficiency	90%
Units/shift	658

$L = 0.95U$

Objective-function value	13
Cycle time	40
Labor cost per unit (seconds)	160
Sub-line efficiency	90%
Units/shift	675

L_1-norm partition

Cycle time	41
Labor cost per unit (seconds)	164
Sub-line efficiency	92%
Units/shift	658

L_2-norm partition

Cycle time	40
Labor cost per unit (seconds)	160
Sub-line efficiency	92%
Units/shift	675

Again, the best results are obtained with $0.95U$ or with the L_2-norm partition. Note that, although the penalty cost is nonzero, the $0.95U$ partition is better than those with a smaller value of L.

Console assembly line

Labor ideal cost per unit (seconds)	124
Number of operators	3
Total working time (seconds)	81000

In this case the same partition was found by all the objective functions, and is depicted in Figure 11.8. It yields:

Cycle time	42
Labor cost per unit (seconds)	126
Sub-line efficiency	98%
Units/shift	642

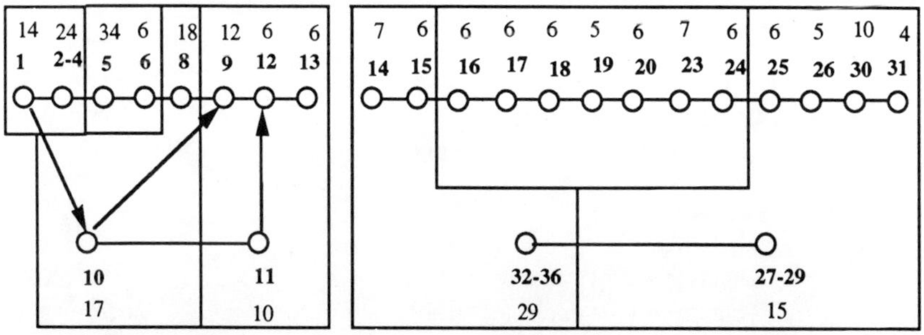

Optimal *L2-norm* partition for shells & boxes assembly line

Optimal *L2* -normpartition for console & radiator assembly line

Figure 11.8

Final heating devices assembly on modules As explained earlier, two situations were analyzed, with two and three operators respectively. Since there are five sub-lines in parallel, for each sub-line the desired cycle time is $43.2 \times 5 = 216$.

Two operators

Labor ideal cost per unit (seconds)	533
Number of operators	2
Total working time (seconds)	270000

In this case also the same partition was found by all the objective functions; it is depicted in Figure 11.9. It yields:

Cycle time (on each sub-line)	269
Labor cost per unit (seconds)	538
Sub-line efficiency	99%
Units/shift (total over the five sub-lines)	501

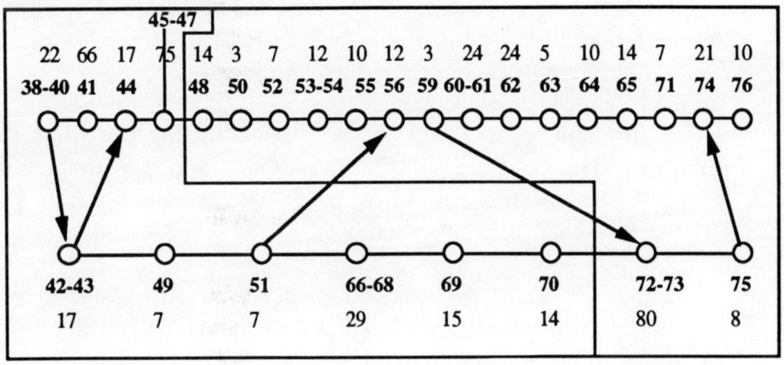

Figure 11.9 Optimal L_2-norm partition for final heating device assembly on modules with 2 operators.

Three operators

Labor ideal cost per unit (seconds)	533
Number of operators	3
Total working time (seconds)	405000

L–U partitions with $C_A = 1$ and $C_B = 10$

$L = 0.7U$

Objective-function value	0
Cycle time (on each sub-line)	206
Labor cost per unit (seconds)	618
Sub-line efficiency	86%
Units/shift (total over the five sub-lines)	655

$L = 0.8U$

Objective-function value	0
Cycle time (on each sub-line)	179
Labor cost per unit (seconds)	537
Sub-line efficiency	99%
Units/shift (total over the five sub-lines)	754

$L = 0.95U$

Objective-function value	82
Cycle time (on each sub-line)	203
Labor cost per unit (seconds)	609
Sub-line efficiency	88%
Units/shift (total over the five sub-lines)	665

L_1-norm partition

Cycle time (on each sub-line)	215
Labor cost per unit (seconds)	645
Sub-line efficiency	83%
Units/shift (total over the five sub-lines)	627

L_2-norm partition

Cycle time (on each sub-line)	179
Labor cost per unit in seconds	537
Sub-line efficiency	99%
Units/shift (total over the five sub-lines)	754

Unlike previous cases, we observe that increasing the value of L gives worse solutions. The L_2-norm solution is the best, and is illustrated in Figure 11.10. Using the results of the L_2-norm partitions for each separate assembly line, we get the following overall values:

Labor ideal cost per unit (seconds)	804
Number of operators	20
Cycle time (global)	42
Units/shift	642
Labor cost per unit	841

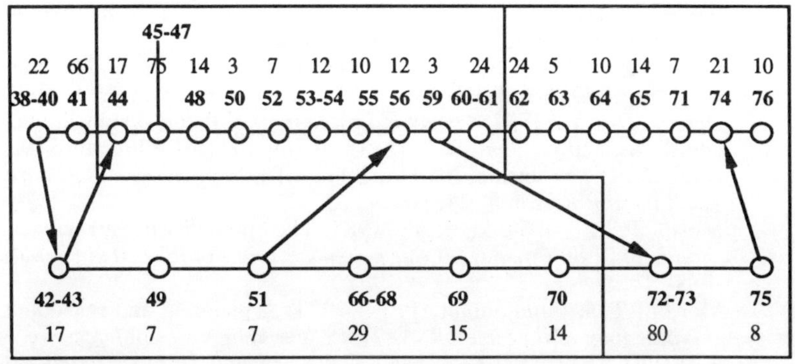

Figure 11.10 Optimal L_2-norm partition for final heating device assembly on modules with 3 operators.

5 CONCLUSIONS

In this chapter the problem of assigning the operations to the workstations in a flexible flow line was considered. A class of graphs, relevant from the application viewpoint, was analyzed; the class was such that the problem could be solved exactly in reasonable time even with small computers. The pipeline structure of the layout plays a major role in determining the complexity of the problem. The same problem can be tackled in a more general context, in which no special assumption is made on the topology of the transportation system: this is what is usually called *loading* problem in flexible manufacturing systems (see for instance, Stecke, 1983; Morton and Smunt, 1984; Chakravarty and Shtub, 1984). In that case, workload balancing is still the main objective (Morin and Stecke, 1985; Schriber and Stecke, 1988; Stecke and Kim, 1989), but the total number of part transfers can significantly affect the quality of the solution (Agnetis *et al.*, 1990), and therefore cannot be disregarded in the optimization process.

The problem here was analyzed by using different objective functions; from a practical viewpoint, the one that was considered most useful by the shop-floor operators, was the L_2-norm objective function, yielding smoothly balanced workloads and low cycle times.

An interesting open problem concerns the possibility of exactly solving the problem by means of algorithms having lower complexity than that discussed in Section 3.2

REFERENCES

Agnetis, A., Arbib, C., Lucertini, M., and Nicolò, F. (1990) Part routing in flexible assembly systems, *IEEE Transactions on Robotics and Automation*, 6(6), 697–705.

Agnetis, A., Arbib, C., Lucertini, M., and Nicolò, F. (1991) *Task Assignment and Subassembly Synchronization in Pipeline Assembly Systems*, Università di Roma 'Tor Vergata', Centro Vito Volterra, Report no. 81.

Askin, R. G., and Standridge, C. R. (1993) *Modeling and Analysis of Manufacturing Systems*, John Wiley & Sons, New York.
Chakravarty, A. K., and Shtub, A. (1984) Selecting parts and loading FMS, *Proceedings of the 1st ORSA–TIMS Special Interest Conference on FMS*, pp. 230–235.
Denardo, E. V. (1982) *Dynamic Programming*, Prentice-Hall, Englewood Cliffs, NJ.
Ghosh, S., and Gagnon, R. J. (1989) A comprehensive literature review and analysis of the design, balancing and scheduling of assembly systems, *International Journal of Production Research*, **27**, No. 4, 637–670.
Morin, T. L., and Stecke, K. E. (1985) The optimality of balancing workloads in certain types of flexible manufacturing systems, *European Journal of Operational Research*, **20**, 68–82.
Morton, T. E., and Smunt, T. L. (1984) A planning and scheduling system for FMS, *Proceedings of the First ORSA–TIMS Special Interest Conference on FMS*, pp. 313–326.
Pizzichella, M. (1989) *Il problema del bilanciamento delle linee per un sistema di assiematura di un'apparecchiatura elettromeccanica per autovettura*, Laurea Thesis, Dipartimento di Informatica e Sistemistica—Università di Roma 'La Sapienza'.
Schriber, T., and Stecke, K. E. (1988) Machine utilizations achieved using balanced flexible manufacturing systems production ratios in a simulated setting, *Annals of Operations Research*, **15**, 229–267.
Stecke, K. E., and Kim, I. (1989) Performance evaluation for systems of pooled machines of unequal sizes: unbalancing versus balancing, *European Journal of Operational Research*, **42**, 22–38.
Stecke, K. E. (1983) Formulation and solution of nonlinear integer production planning problems in FMS, *Management Science*, **29**, No. 3, 273–288.

Alessandro Agnetis, Alessandro Ciancimino and **Massimo Pizzichella**
Dipartimento di Informatica e Sistemistica
Università di Roma 'La Sapienza'
Via Buonarroti 12
00185 Rome
Italy

Mario Lucertini
Dipartimento di Ingeneria Elettronica
II Università di Roma
Via della Ricerca Scientifica
00133 Rome
Italy

12
Optimization Algorithms for Improving Cam Lobes toward Reduced Valve-Train Vibrations in Automotive Engines

R. Ernst and J. Meyer
Ford Motor Company, Köln, Germany

ABSTRACT
Two optimization methods for improving valve-train behavior are presented. To do this the following aspects are of importance:

(i) accurate prediction methods for calculating the dynamic behavior of the mechanism,
(ii) analytical tools to quantify the performance of the valve train,
(iii) well-defined design parameters,
(iv) efficient optimization algorithms.

The methods can be used both as a design tool and for gaining insight into the physics of the mechanism. Numerical results indicate a stability benefit from adjusting the cam-deceleration phase according to force oscillations of the valve spring.

Further, it is shown on the example of a low-frequency push-rod valve train, that it is possible to significantly reduce the vibration excitation by changing the contour of the first acceleration peak of the cam to a double peak.

1 INTRODUCTION

The layout of valve-lift curves is of major importance for many aspects of automotive engines. The requirements of a valve train in high revving combustion engines come from the following areas. To achieve optimal gas exchange the valve

Optimization in Industry 2. Edited by T. A. Ciriani and R. C. Leachman

should open in a very short time. This correlates with large accelerations and decelerations. On the other hand, to satisfy high-durability demands one needs to keep the mechanical loads as small as possible. And, third, to keep NVH behavior (noise vibration harshness) within limits mechanical impacts should best be eliminated, corresponding to a good control of the valve-train movement.

Therefore, valve-lift design is a compromise where the conflicting restrictions can be very tight. To find the design which best meets the requirements it seems sensible to quantify the conflicting arguments and to balance them by changing design parameters and looking at the numerical values describing various qualities. The following sections describe possible procedures. In Section 2, methods for calculating the kinematic and dynamic behavior of valve trains are described. Section 3 gives an overview on the ways of formulating objectives for parameter optimization. The parameter-optimization methods are referred to as direct methods. In Section 4, results of a calculation run are demonstrated and discussed for the example of a bucket-tappet valve train. In Section 5, optimal control theory is briefly outlined and discussed. This method will be referred to as an indirect method. In Section 6, the indirect method is applied to vibration reduction in a push-rod valve train. The chapter closes with a summary and outlook.

2 KINEMATICS AND DYNAMICS

Numerical analysis of the dynamic behavior of valve trains has been under study for many years (Barkan, 1953; Dudley, 1948). In fact, knowledge in the field is still evolving (Bakaj, 1988; Phlips *et al.*, 1989; Phlips and Schamel, 1991; Seidlitz, 1990). In the case of direct methods the valve-lift curve is calculated by defining the corresponding valve acceleration via parametric splines of the third order. During the optimization process these spline parameters will be used. After having defined the valve-acceleration curve lift and velocity can be calculated by integration.

In most applications a dynamic model of the valve train under study is formulated as a system of coupled elastic and rigid bodies as shown in Figure 12.1. The rigid bodies are connected by springs and dampers. The dynamics of elastic bodies such as the valve spring can be accounted for either by including the lowest few eigenmodes or by discretizing the elastic body into coupled rigid bodies. Thus the dynamics is formulated as a coupled set of differential equations. Solving these equations to a high accuracy has proved to be a demanding task. The physical effects to be included in the model depend on the aims of the analysis. If the aim is to predict at exactly what speed contact loss within the valve-train parts has to be encountered, then one needs a very accurate knowledge of the stiffnesses within the valve train. In a bucket-tappet valve train this would be the stiffness of the bucket which varies greatly as a function of the point of cam-to-tappet contact. FEM calculation can therefore be of great help on this issue. If the valve train contains a hydraulic lash adjuster one has to account for this variable stiffness (Phlips *et al.*, 1989).

Parameter-optimization algorithms typically use a very large number of evaluations of the dynamic equations. Because of this it is of importance to be

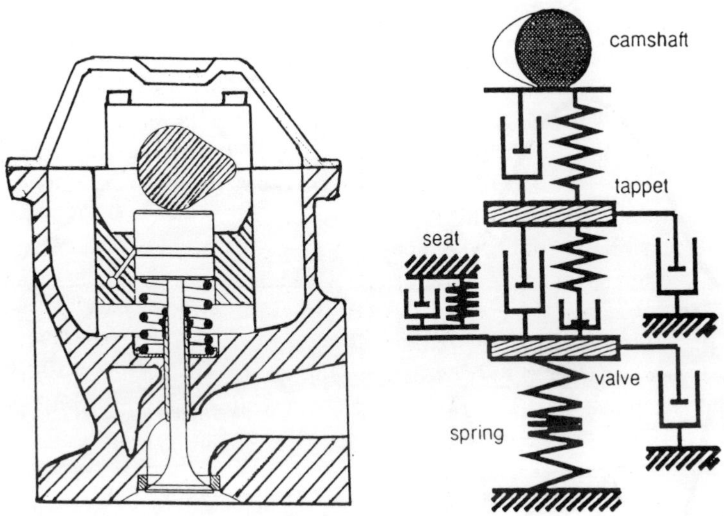

Figure 12.1 (a) Sketch of a bucket tappet valve train (figure INA). (b) Dynamic model for a bucket tappet valve train.

able to solve the equations with a minimum of computational effort and still maintain high accuracy. Here it is shown that solving only for the difference of the elongations between neighboring components gives a great reduction in computer time (Phlips *et al.*, 1989).

In the following discussion, a bucket-tappet valve train will be used to demonstrate the method. Figure 12.2 shows the kinematic valve lift, the valve velocity and the valve acceleration suited for such a valve train. Figure 12.3 depicts the dynamic behavior of several observables of the valve train for a typical cam lobe.

Figure 12.3a shows the dynamic-valve acceleration for two different speeds called 'high speed' and 'low speed'. The low-speed curve shows oscillatory behavior with one dominating frequency; high frequencies only occur at the end of the event where the valve gets into contact with the valve seat again. The high-speed curve has additional features. The most important is a strong change in acceleration at a cam angle of 80°. This change in acceleration indicates an impact which occurs after loss of contact between cam and tappet. Impacts typically have NVH problems. Another strong impact can be seen slightly before the end of the event, indicating early seating. If this happens there is no longer control over the speed and force with which the valve hits the valve seat. Such a behavior can destroy valves and thus be fatal for the engine.

In Figure 12.3b the force between cam and tappet is depicted for the same speeds as in Figure 12.3a. If the force is equal to zero there is contact loss over this cam-angle interval. The angle at which contact is gained again (80°) correlates with the impact and with the sudden change in dynamic acceleration at the valve. After the event one can still see intervals of contact loss, indicating valve bouncing.

In Figure 12.3c the dynamic valve-spring force is shown. Oscillations of the spring force resulting from the inner oscillations of the valve spring can be seen. For

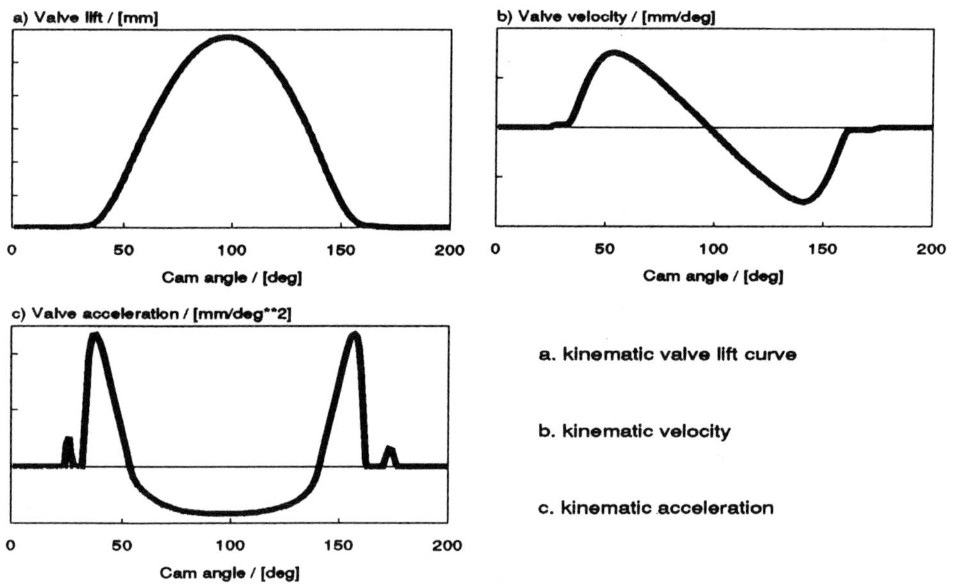

Figure 12.2 Kinematic movement of a valve.

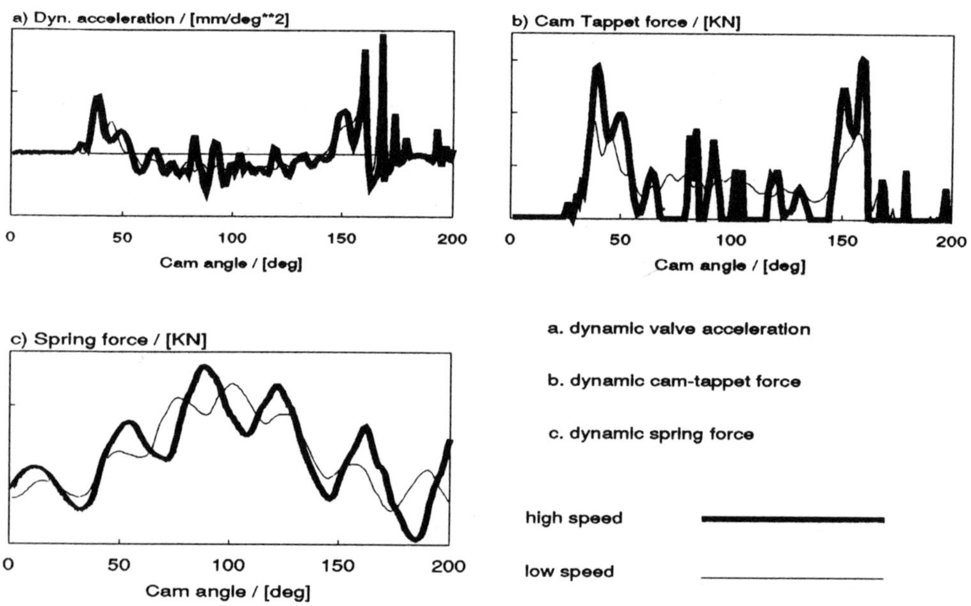

Figure 12.3 Dynamic valve acceleration, cam-tappet force and spring force for two different speeds.

the high-speed case the amplitude of the spring oscillation gets very large. At the minimum force around cam angle 70° to 80° the spring exerts around 300 N less force than needed. This is exactly where we find contact loss occurring for the first time. All the other intervals of contact loss occur during dynamic spring-force minima.

3 DIRECT METHOD

As can be seen from the previous section arguments for a good dynamic valve-train behavior are:

1 reduction of oscillation amplitudes as much as possible;
2 keeping contact between the parts of the valve train within a given speed range;
3 if possible increase the area under the lift curve to improve gas exchange;
4 reduction of the speed at which parts come into contact, e.g. during the valve-seating process.

In addition to the criteria 1–4 several further boundary conditions need to be satisfied. Typical criteria for a bucket-tappet valve train are:

5 The radius of curvature of the cam has to be kept within limits. It should not get too small at the cam tip because of durability arguments.
6 Differential geometry shows that the location of contact between cam and tappet is a function of the derivative of the valve-lift profile. In order to prevent the contact point going further than the radius of the tappet allows, the derivative of the valve-lift curve has to be kept below a maximum value.
7 Using splines for defining the second derivative of the profile, we have to take care that the first derivative and the valve-lift contour itself reach the desired ramp values at the end of the event.

The formulation of the optimization goals 1–3 and restrictions 5–7 can be accomplished by the following measures:

An example for a measure of oscillations can be

$$F_1 = \int_{\text{event}} (A_{\text{dyn}} - A_{\text{kin}})^2 \mathrm{d}t.$$

The measure for the amount of contact loss is

$$F_2 = \sum_{I_j} \mu^2_{(I_j)}$$

The measure for the area under the lift curve is

$$F_3 = \int_{\text{event}} (L - L_{max})\mathrm{d}t.$$

A_{dyn} = dynamic acceleration of the valve,
A_{kin} = kinematic acceleration of the valve,
I_j = interval j at which contact loss appears,
$\mu_{(I_j)}$ = length of interval I_j,
L = valve lift,
L_{max} = maximal valve lift.

The constraints can be included by the penalty-function approach. It states that any cam profile is checked for, whether a constraint is violated or not. If one or more are violated, the penalty function would be the squared amount of the violation. If none are violated the penalty function takes the value zero. For each of the above restrictions a penalty function is formulated. The total objective function is then calculated by simply adding all the terms. Each term can be multiplied by a weighting factor balancing the different arguments. The constraints will be weighted with a very large number, because these have to be satisfied without compromise.

The rest of the arguments of the objective function can be weighted according to engineering requirements.

The total objective function can then be formulated as follows:

$$F_{tot} = \sum_{i=1}^{3} F_i + \sum_j R_j$$

where

F_{tot} = total value of the objective function,
F_i, $i = 1, \ldots, 3$ = the values of the optimization goals in this example,
R_j = values of restriction functions describing the constraints.

After the objective function is defined an optimization routine using Powell's algorithm (Press *et al.*, 1989) is taken to change the acceleration-profile parameters in order to find a minimum of the target function. The minimum search is accomplished by a series of line minimizations.

4 RESULTS OF DIRECT METHOD

In the calculation demonstrated the valve curves from Figure 12.2 were used as a start which was then iteratively improved by the optimization routine. The dynamic equations were solved for ten different speeds from 3800 crank-shaft

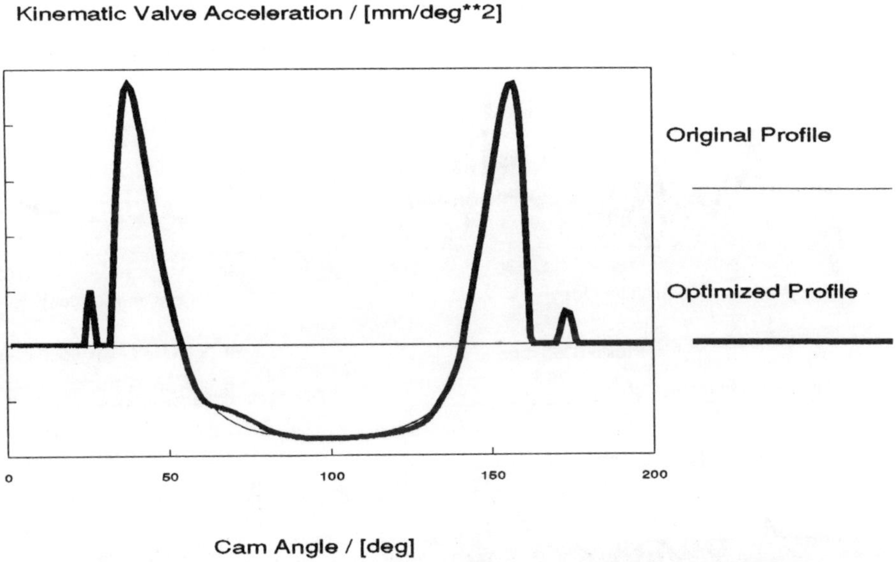

Figure 12.4 Kinematics of original valve profile and optimized profile. (Bucket tappet valve train).

revolutions per minute up to 6200 r.p.m. The aim was to reduce the length of intervals with contact loss over the given speed range and to reduce the amplitudes of the acceleration oscillations. In Figure 12.4 the accelerations of the old profile and the improved profile are depicted.

The major effect that can be seen is in the deceleration phase. The angle at which the value of the deceleration was reduced in the optimized profile fits with the minimum force as a result of the valve-spring oscillation. The physical meaning of this is that with the deceleration the inertial forces get smaller, so that the valve spring has to hold only as much mass forces as it actually can. This results in less contact loss over the event and speed range. Figure 12.5 shows the dynamic valve acceleration for the original cam and the optimized cam for two different speeds. It can be seen that the oscillation amplitudes for the optimized profiles get smaller, and the peaks at the end of the event are reduced, which indicates a smaller impact force for the valve-seating process.

In Figure 12.6 the force between cam and tappet is depicted for the same calculation. Here it can be seen that the amount of contact loss is significantly reduced. Over the whole optimized speed range it was possible to reduce the amount of contact loss by 70% and the difference between dynamic and kinematic acceleration by 45%, which is a significant improvement.

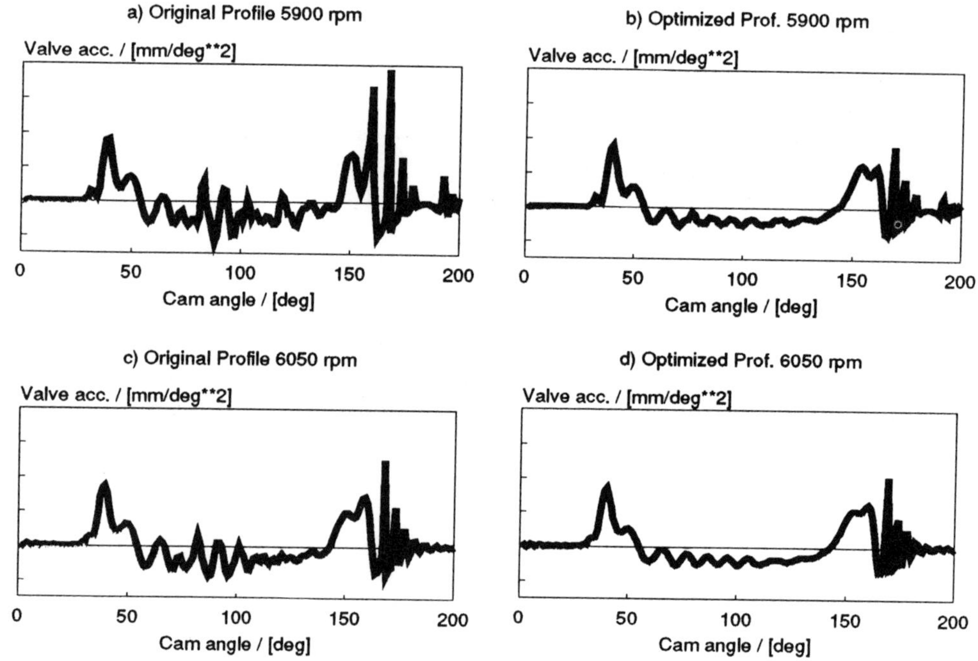

Figure 12.5

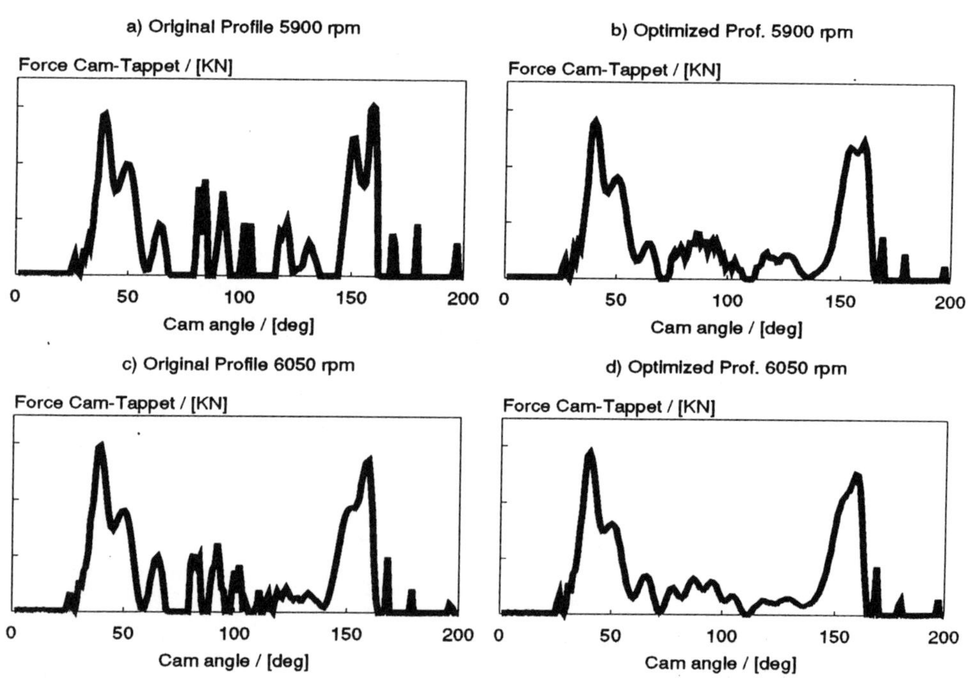

Figure 12.6

5 INDIRECT METHOD

Indirect methods reformulate the original optimization problem as a multi-point boundary-value problem (Bryson and Ho, 1969). To do this it is of help to first describe the optimization problem in a standard form.

Given a dynamic system described by equations (1) with initial conditions (2),

$$\mathrm{d}\vec{x}/\mathrm{d}t = \vec{f}(\vec{x}(t), u(t)), \tag{1}$$

$$\vec{x}(0) = \vec{x}_0, \tag{2}$$

one wants to minimize the functional J described by equation (3). The function u on the right-hand side of (1) is called the control function.

$$J = \Phi(\vec{x}(T)) + \int_0^T l(\vec{x}(t), u(t))\mathrm{d}t. \tag{3}$$

The optimal control is that u which minimizes J. It can be shown that the optimal control u satisfies the minimum principle which is formulated by equations (4):

$$\mathrm{d}\vec{x}/\mathrm{d}t = \vec{f}(\vec{x}(t), u(t)), \tag{4a}$$

$$\vec{x}(0) = \vec{x}_0, \tag{4b}$$

$$\begin{aligned} -\mathrm{d}\vec{\lambda}(t)/\mathrm{d}t^T &= \vec{\lambda}(t)^{\mathrm{T}}\vec{f}_{\vec{x}}(\vec{x}(t), u(t)) + \mathrm{l}_{\vec{x}}(\vec{x}(t), u(t)) \\ &= H_{\vec{x}}(\vec{x}(t), u(t)) \end{aligned} \tag{4c}$$

$$\vec{\lambda}(T)^T = \Phi_{\vec{x}}(\vec{x}(T)). \tag{4d}$$

For all t, $0 \leqslant t \leqslant T$, and for all possible controls

$$H(\vec{\lambda}(t), \vec{x}(t), v) \geqslant H(\vec{\lambda}(t), \vec{x}(t), u(t)). \tag{4e}$$

Equations (4) represent a boundary-value problem. Its solution is a candidate for an optimal control. These equations have been adapted to a simple one-mass valve-train model, with the aim of minimizing the compression of the valve train.

6 RESULTS OF INDIRECT METHOD

Results are presented for the improvement of a V6 engine with a push-rod valve train. Figure 12.7 shows the corresponding lumped-mass model (Festag, 1989). In Festag (1989) one can also find correlation data between

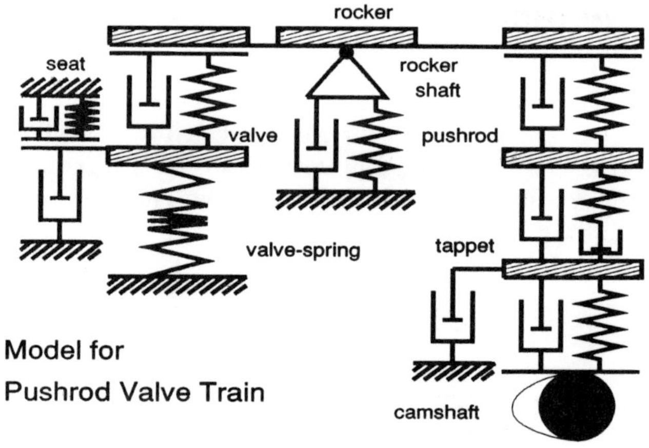

Figure 12.7

measurements and calculation of the dynamic valve acceleration. The lowest eigenfrequency of this valve train is around 600 Hz. This is a very low value which results in strong oscillations of the mechanism. The optimization aim was to reduce oscillations, while maintaining a cam-lift curve close to a standard cam-lift curve. Because of this, only the first section of the cam was optimized with the condition that lift, velocity and acceleration fit steadily to the unchanged part of the profile. In Figure 12.8 the original acceleration profile is depicted together with the

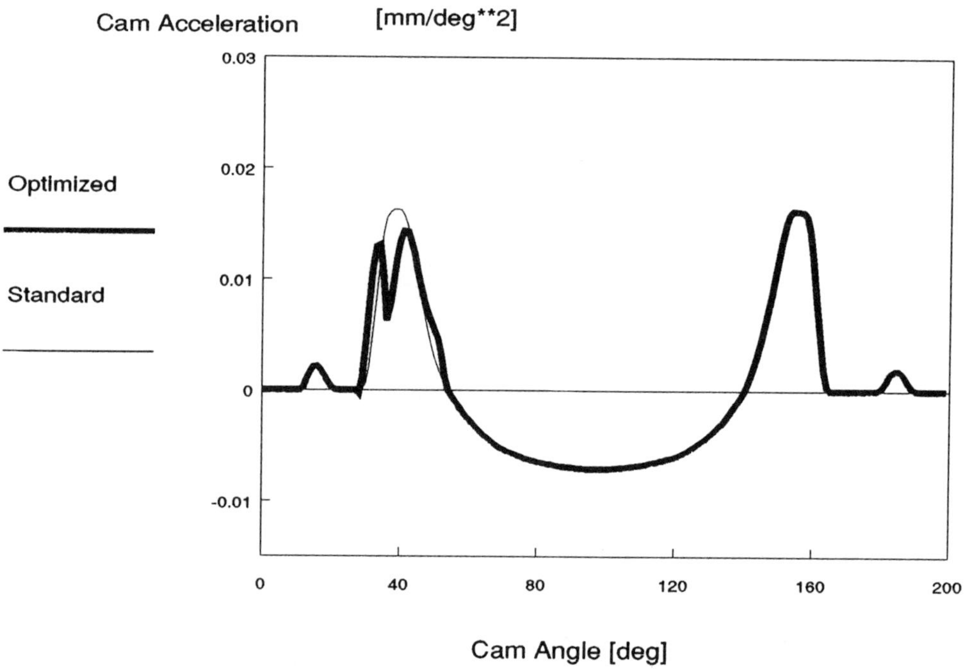

Figure 12.8

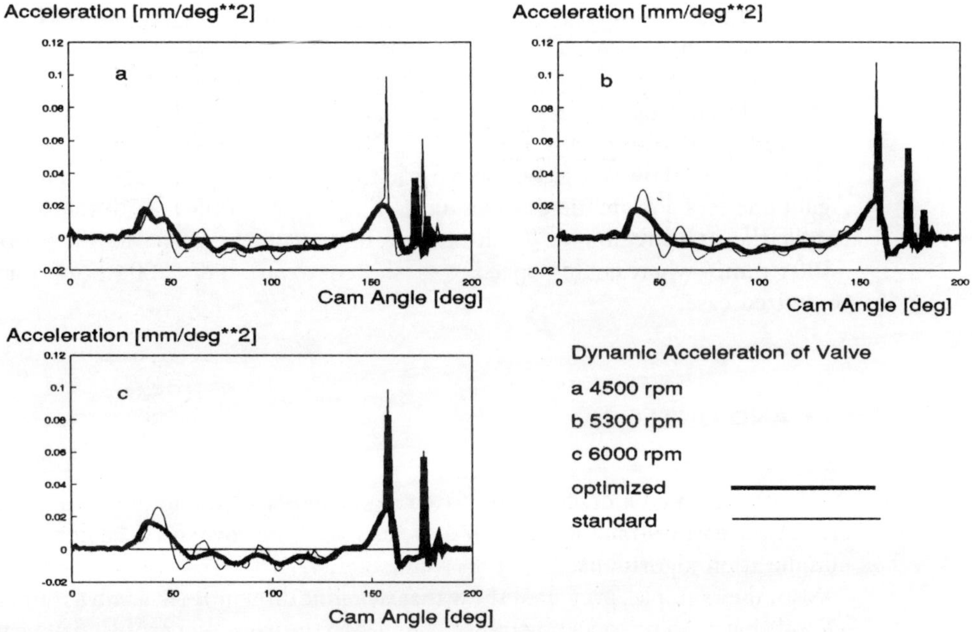

Figure 12.9

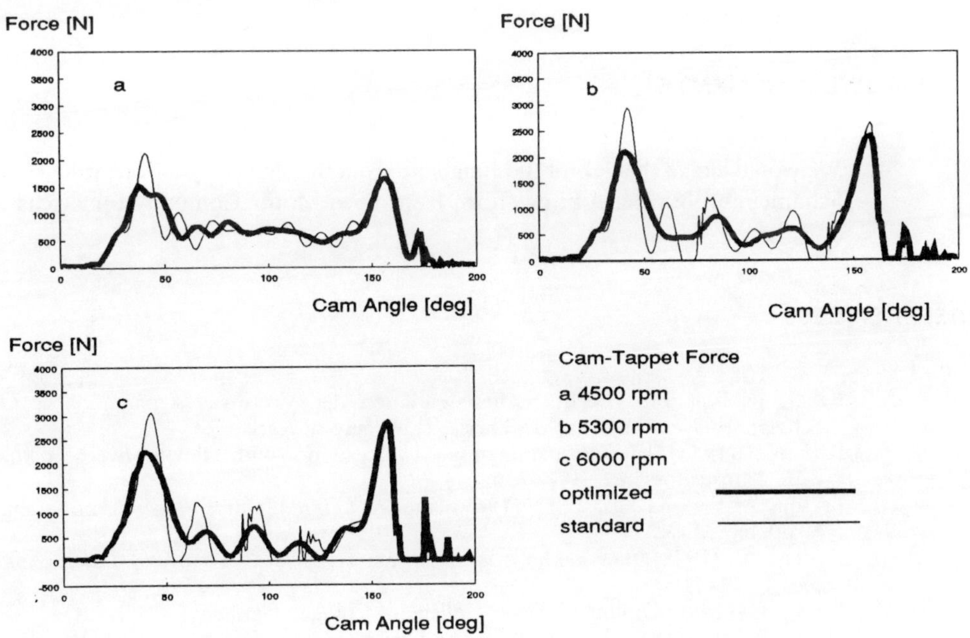

Figure 12.10

optimized one. The first acceleration peak has changed to a double peak. It can also be seen that from 55° on, the two cams are identical.

In Figure 12.9 the dynamic valve accelerations of the two cams are compared at three different speeds. It can be seen that in the optimized case the oscillation amplitudes are significantly reduced.

Figure 12.10 shows the same comparison for the force between cam and follower. Again one sees a dramatic decrease in oscillation amplitudes. With the standard profile the speed region at which one has to encounter contact loss lies around 5300 r.p.m., whereas this speed was shifted to around 6000 r.p.m. for the optimized case.

7 SUMMARY AND OUTLOOK

We have presented a description of two algorithms which allow one to quantify and improve valve-train behavior. It was possible to improve this behavior by using minimization algorithms.

Also, the examples presented show that dynamic difficulties in a valve train have a good chance to be solved by minor changes to the cam-lobe profile, rather than the traditionally used method of increasing the valve-spring force, which in itself leads to the adverse consequence of increased friction.

For further work it seems reasonable to formulate criteria to account for various engine states such as different amount of oil aeration etc.

ACKNOWLEDGEMENTS

We would like to thank Prof. G. Jungbluth from the University of Karlsruhe, Andreas Schamel and Burkhard Pfingsthorn from Ford Motor Company for discussions.

REFERENCES

Bakaj, L. F. (1988) Rechnerische Simulation der Ventiltriebsdynamik eines OHC-Schlepphebelventiltriebs, PhD Thesis, University of Karlsrule.

Barkan, P. (1953) Calculation of high speed valve motion with a flexible overhead linkage, *SAE Transactions*, **61**, 687–700.

Bryson, A. E., and Ho, Y.-C. (1969) *Applied Optimal Control*, Ginn and Company, Waltham, Mass.

Dudley, W. (1948) New methods in valve cam design, *SAE Quarterly Transactions*, **2**, No. 1, 19–51.

Festag, G. (1989) Diploma Thesis, College of Cologne Germany.

Phlips, P., and Schamel, A. (1991) *The Dynamics of Valvetrains with Hydraulic Lash Adjusters and the Interaction with the Gas Exchange Process*, SAE Technical Paper 910071.

Phlips, P., Schamel, A., and Meyer, J. (1989) *An Efficient Model for Valvetrain and Spring Dynamics*, SAE Technical Paper 890619.

Press, W. H., Flannery, B. P., Teukolsky, S. A., and Vetterling, W. T. (1989) *Numerical Recipes: The Art of Scientific Computing*, Cambridge University Press.

Seidlitz, S. (1990) *An Optimization Approach to Valve Train Design*, SAE Technical Paper 901638.

Ford Motor Company
John-Andrews-Entwicklungszentrum
Dept. MD/PN-320
5000 Köln
Germany

13
On Modeling Robust Policies for Financial Trading

S. Ahn[1], L. F. Escudero[2], and M. Guignard-Spielberg[1]
[1]*University of Pennsylvania, PA, USA*
[2]*Universidad Complutense de Madrid, Spain*

ABSTRACT
Several mixed 0–1 programs for trading financial instruments with uncertainty in interest rates along a planning horizon are proposed. In contrast to traditional approaches, we use scenario groups to characterize the uncertainty. Solutions (i.e. trading policies) are obtained for each scenario group and, then, these individual solutions are aggregated to yield a nonanticipative or implementable policy that minimizes the regret of wrong decisions. The regret of a wrong decision is defined as the expected lower partial difference between the net return of the proposed trading policy and the net return of the optimal trading under any scenario group at the end of the planning horizon.

1 TRADING FINANCIAL INSTRUMENTS

Let us assume a (short-term) planning horizon (e.g. 3 or 6 months) with (integer) time periods. The time periods may have unequal length (e.g. days, weeks, months, quarters). A set of financial instruments is available for trading. The trading value of a financial instrument at a time period depends upon the number of periods up to the maturity of the instrument, the interest-rate path up to the maturity period, the return of the investment at given periods, and a weighting factor that takes into account the specifics of the instrument.

Given the trading value for each instrument and the cash availability for each period of the planning horizon, the goal consists of obtaining the volume to buy and sell for each instrument to allow certain trading positions along the planning horizon that satisfy also some logical constraints. The objective is to maximize the investor's total wealth at the end of the planning horizon.

Optimization in Industry 2. Edited by T. A. Ciriani and R. C. Leachman

A right decision requires appropriate information about the available financial instruments and their trading value. Assuming all data is known with certainty, linear and mixed 0–1 programs can be designed and state-of-the-art software can be used. Nevertheless, the interest-rate path along the planning horizon is not always known in advance and, then, traditional approaches are not valid. Instead, we need to model the uncertainty in a domain where, very frequently, historical data is not enough. On the other hand, the criteria to define the goodness of a solution in the deterministic field cannot be applicable in a domain where different scenarios can occur and none of them is under the control of the modeler. Then, the notion of optimality should be replaced in the stochastic field by some other concept where the uncertainty is taken into account.

Given a set of potential scanarios for the interest-rate path and related weight, we use the notion of scenario immunization to compare one solution with another. The goal, then, is to obtain a decision policy that minimizes the expected difference between the terminal wealth due to the proposed policy and the optimal terminal wealth under any scenario that may occur. In this way, the modeler hedges himself against the volatility of the interest rate and the disastrous consequences if some extreme scenario in fact happens.

We introduce the notion of scenario group for a given time period, such that two scenarios belong to the same scenario group if they are identical up to that period on the basis of the information available about them at that period (i.e. the interest rate). Based on the implementable principle introduced by Rockafellar and Wets (1987), the decision variables for a given time period should be identical for all scenarios that belong to a given group at that period.

A decision policy is termed *robust* if it guarantees scenario immunization and, additionally, it preserves the scenario-group implementable principle.

The chapter is organized as follows. Section 2 gives the deterministic program. Section 3 provides an introduction to the principles for our stochastic approach. Our scenario-immunization methodology is given in Section 4. The model for obtaining robust policies is presented in Section 5. It results in a huge program whose dualization is dealt with in Section 6. Finally, we draw some conclusions.

2 MIXED 0–1 PROGRAM

The (deterministic) program assumes that the interest rate is known by the market for each period of the planning horizon. Before introducing the program, let us give the notation.

Data

T, number of periods in the planning horizon (t).
I, set of financial instruments (i).
J, set of types of financial instruments (j).
I_j, set of financial instruments that belong to type j, such that $I_j \subseteq I$, $I_j \cap I_g = 0$, for $j,g \in J$.

K_i, issuing period for instrument i, where $0 \leqslant K_i < T_i$. (Note: by convention it is assumed $K_i = 0$ for the case where the instrument was issued prior to any period in the planning horizon.)

T_i, maturity period for instrument i for $i \in I$, $1 \leqslant T_i$.

T^j, maturity period for instrument type j, such that

$$T^j = \max_{i \in I_j} \{T_i\}.$$

$C_{i\tau}$, yield of instrument i at (the end of) period τ for $K_i < \tau < T_i$. (Note: for discount and zero-bonds, $C_{i\tau} = 0$ for $\tau < T_i$.)

r_t, (discount) interest rate for period t for $0 < t < \max\{T, \max_i\{T_i\}\}$.

f_j, weighting factor on the interest rate for computing the value of the instruments in set $I_j, j \in J$. See below.

$E_{jt\tau}$, updated value at (the end of) period t for the unit yield of any instrument of type j at (the end of) period τ. Although in this work we assume that the unit value is constant, we can easily broaden our approach to the case where the unit value is (possibly) a piecewise linear function of the trading volume.

$$E_{jt\tau} = \Pi_{t<l<\tau}(1 + f_j \tau_l)^{-1}, \quad j \in J, \quad 0 < t < \tau \leqslant T, \quad t \leqslant T, \tag{1}$$

Note that without loss of generality, we can assume that all instruments for a given type and same period for return of investment, say τ, have the same unit-value path from any period, say t, up to the given period τ. For instance, 1, 3, and 6 months. Treasury bonds with identical periods for return of investment have the same unit value at any period t, since the solvency and liquidity are identical for all of them.

P_{it}, unit market price of instrument i at (the end of) period t. It can be expressed as

$$P_{it} = \sum_{t<\tau\leqslant T_i} C_{i\tau} \mathrm{E}_{\mathrm{jt}\tau}, \quad i \in I_j, \quad j \in J, \quad 0 < t \leqslant T_i. \tag{2}$$

Note. P_{i0} is the unit market price in spot. The weighting factor has been used in (1) as an *option adjusted premium* (OAP), see Zenios (1992). The OAP methodology estimates the factor for the interest-rate path that will minimize the weighted difference between the (observed) market price $P_{i0} \forall i \in I_j$ and the 'fair' price obtained by applying the expectations hypothesis based on (1) and (2) for all instruments in I_j. (As an alternative to $f_j \tau_l$, where f_j is used as a *multiplicative* OAP, the additive $f_j + \tau_l$ can be used, see Zenios, 1991.)

z_{i0}, initial stock for instrument i.

D_t, exogenous cash availability at (the end of) period t for $0 \leqslant t < T$. (Note: it is assumed that the cash amount is available for any period prior to the trading session.)

N_t, maximum number of instrument types to hold in stock at period t for $0 < t \leqslant T$.

F_i, unit face value of instrument i for $i \in I$.

m_{jt} and M_{jt}, conditional minimum and maximum (face) value of instrument type j that is allowed to hold at period t, for $j \in J$, $0 < t \leqslant T$.

$\overline{X}_{it}^{+}$ and $\overline{X}_{it}^{-}$, maximum allowed volume of instrument i to be bought and sold, respectively, at (the end of) period t, for $0 \leqslant t < T$, $i \in I$.

v_j, transaction cost per unit face value of instrument type j.

$P_{it}^{-} = P_{it} - v_j$, unit revenue of instrument i sold at (the end of) period t for $0 < t < T_i$, $i \in I_j$, $j \in J$.

$P_{it}^{+} = P_{it} + v_j$, unit cost of instrument i bought at (the end of) period t for $0 < t < T_i$, $i \in I_j$, $j \in J$.

Variables

X_{it}^{+} and X_{it}^{-}, volume of instrument i to be bought and sold, respectively at (the end of) period t for $0 \leqslant t < T$, $i \in I$.

Z_{it}, stock (volume) of instrument i at (the beginning of) period t for $0 < t \leqslant T_i$.

δ_{jt}, a 0–1 variable such that its value is 1 if stock of financial instrument type j is held at period t and, otherwise, the value is zero.

S_t, net cash at (the end of) period t after trading the instruments and cashing their returns for $0 \leqslant t \leqslant T$.

The goal is to obtain the trading volume (X^{+}, X^{-}) and, then, the instrument's stock Z and net cash S along the planning horizon to maximize the net cash and the stock trading value at the end of the planning horizon (the so-called terminal wealth), such that (i) the exogenous cash availability per period is not violated, (ii) the trading volume per instrument and period does not exceed the allowed value, (iii) the (face) value to hold for each instrument type is within allowed limits per period, if any, and (iv) the maximum number of instruments to hold at any time is not violated.

The following instrumental parameter makes easier the representation of the model:

ρ_{it}, index with value 1 for trading periods of instrument i and, otherwise, it is zero for $i \in I$ and $0 \leqslant t \leqslant T$, this is

$$\rho_{it} \equiv \begin{cases} 1 & \text{for } K_i \leqslant t < T_i, \\ 0, & \text{otherwise.} \end{cases} \tag{3}$$

The mixed 0–1 program is as follows:

$$V(*) = \max S_T + \sum_{i \in I} \rho_{iT} P_{iT}^{-} Z_{iT}, \tag{4}$$

subject to

$$Z_{it} + X_{it}^{+} - X_{it}^{-} = Z_{i,t+1}, \quad i \in I, \quad 0 \leqslant t < T, \tag{5}$$

$$-S_{t-1} + \sum_{i \in I} [-C_{it} Z_{it} + \rho_{it} P_{it}^{+} X_{it}^{+} - \rho_{it} P_{it}^{-} X_{it}^{-}] + S_t = D_t, \quad 0 \leqslant t < T, \tag{6}$$

$$S_{T-1} + \sum_{i \in I} C_{iT} Z_{iT} = S_T, \tag{7}$$

$$m_{jt} \delta_{jt} \leqslant \sum_{i \in I_j} F_i Z_{it} \leqslant M_{jt} \delta_{jt}, \quad j \in J, \quad 0 < t \leqslant T, \tag{8}$$

$$\sum_{j\in J} \delta_{jt} \leqslant N_t, \quad 0<t\leqslant T, \tag{9}$$

$$0\leqslant X_{it}^{+}\leqslant \overline{X}_{it}^{+}, \quad 0\leqslant X_{it}^{-}\leqslant \overline{X}_{it}^{-}, \quad i\in I, \quad k_i\leqslant t<T, \tag{10}$$

$$\delta_{jt}\in\{0;\ 1\}, \quad j\in J, \quad 0<t\leqslant T, \tag{11}$$

$$0\leqslant Z_{it}, \quad i\in I, \quad K_i\leqslant t\leqslant T_i, \tag{12}$$

$$0\leqslant S_t, \quad 0\leqslant t\leqslant T. \tag{13}$$

The function (4) maximizes the terminal wealth. Constraint (5) is the balance equation for the stock of the financial instruments. Constraint (6) is the balance equation for the cash flow at any period but the last one. (We set $S_{-1}=0$.) Constraint (7) is the balance equation for the cash flow at the last period of the planning horizon. Constraint (8) forces the stock of instrument type j to lie between allowed values, if any. (Note: it is well known that the smaller the difference $M_{jt}-m_{jt}$, the tighter is the LP relaxation of the 0–1 program.) Constraint (9) prevents us holding more financial instruments than the allowed maximum at any time. The bounds on the variables are given in (10)–(13). The model has roughly $T(2+|I|+2|J|)$ constraints, $T(1+3|I|)$ continuous variables, and $|J|\times|T|$ 0–1 variables.

3 STOCHASTIC MODELS FOR FINANCIAL TRADING

Usually, the unit price P_{it} for $t>K_i$ is not known, due mainly to the variability of the interest rate over time; then, one needs to consider two additional features. First, one must model the availability of information over time, and state what sort of decisions can be made at each of the various stages. Second, in order to optimize a poposed solution must be compared with another. But, in the stochastic setting (i.e. a setting where some elements cannot be controlled by the modeler), the criteria by which the judgement can be performed is much less clear. We demonstrate in this section how multi-stage stochastic programs (SP) can be designed to model the information availability, and show that decisions produced by these programs are optimal with respect to the zero-regret criterion—that which could be achieved if the course of the future value of the instruments can be perfectly predicted. Obviously, the 0–1 program (4)–(13) is optimal with respect to the zero-regret criterion, since the interest rate is assumed to be perfectly predicted over time.

In multi-stage decision problems, the decisions made in each period should take into account all future uncertainty and future decisions. We thus need to model the future decision process and information access as well. The specific form of the future decisions depends on assumptions concerning the information that is available to the decision maker, when (in time) it is available and what adjustments (recourse) are available to the decision maker. Varying assumptions regarding the possible adjustments that the decision maker can make lead to different models.

We need an approach to modeling the uncertainty in the problem data. The traditional approach is to make distributional assumptions, estimate the parameters from historic data and, then, develop SP models to take the uncertainty into account, see e.g. Kusy and Ziemba (1986). Such an approach may not be appropriate if only limited information is available. Also, in applications such as forecasting of attributes for financial instruments, it is often necessary and possible to take into account information that is not reflected in the historic data. In many such cases, we may employ a technique called 'scenario analysis', wherein the uncertainty is modeled via a few scenarios, say a set Q.

For example, in our instance, the scenarios would specify the interest rate for each period and, thus, we would have available:

$E^q_{jt\tau}$, updated value at period t for the unit yield of any instrument of type j at period τ under scenario q for $q \in Q$, $0<t<\tau\leqslant T^j$, $j\in I$ and $t\leqslant T$.
w_q, weight (probability) associated with scenario q.

The weights represent the likelihood that the decision maker (modeler) associates with each scenario. (Note: in this work we assume that the exogenous cash availability D_t is estimated with enough accuracy.)

Let r^q_t denote the interest rate for time period t under scenario q for $q\in Q$ and

$$0<t\leqslant \max\{T, \max_i\{T_i\}\}.$$

It is outside the scope of this work to illustrate the details of how these scenarios are generated. (We point out the correlation between the rates r^q_t and r^q_{t+1}.) So, $E^q_{jt\tau}$ can be expressed as

$$E^q_{jt\tau} = \Pi_{t<l\leqslant\tau}(1+f_j\, r^q_l)^{-1}. \tag{14}$$

and, then, the unit market price of financial instrument i at period t under scenario q (say, P^q_{it}) will be

$$P^q_{it} = \sum_{t<\tau\leqslant T_i} C_{i\tau} E^q_{jt\tau}. \tag{15}$$

where $i\in I_j$ for $0<t\leqslant T_i$. Note: $P^q_{it}=C_{it}$ for $t=T_i$, $\forall q\in Q$, $P^q_{i0}=P^{q'}_{i0}\ \forall q,\ q'\in Q$ since it is the market price in spot. (It is assumed that the market has the same information as the modeler.)

One way to use the scenario information is to employ the average value (for each time period) across all the scenarios as an estimate of the value for the instruments at that period but, obviously, such an approach ignores the variability information available in the data. Anyway, let

$$\bar{P}_{it} = \sum_{q\in Q} w_q P^q_{it}. \tag{16}$$

One can solve the 0–1 program (4)–(13) by setting $P_{it} := \bar{P}_{it}$. Another possibility is to solve the 0–1 program for each scenario $q\in Q$, and obtain what can be termed individual scenario solutions. Let $V_q(*)$ denote the optimal objective function value $V(*)$ (4) under scenario q, where P_{it} is replaced by P^q_{it}. The decision maker must then analyze the individual scenario solutions and arrive at a single solution that he can implement.

Although such a procedure is often used in practice with a few scenarios, it can be very tedious and almost impossible when there is a large number of scenarios and time periods. Besides, there is no prescription as to how these solutions are to be analyzed and aggregated. It is clear that the individual scenario solutions are no help to the decision maker, since they presume 'perfect foresight'. Decisions that depend on the individual scenarios do not hedge against the possibility that a given scenario may not occur, leaving one vulnerable to disastrous consequences if some other scenario does in fact happen.

Nevertheless, the scenario optimal solutions are an important starting point for the modeling of decisions under uncertainty. If we had been sure that scenario q is what is going to happen, we could then implement the scenario optimal solution corresponding to q, provided that the market prices the financial instruments as we do (otherwise, see below). Taking such a decision represents an unattainable ideal, which we term zero-regret. The regret is the difference between the optimal value of a candidate trading policy, say $y=(X^+, X^-)$ and the scenario optimal value $V_q(*)$. Regret, in other words, is the amount of money that is left on the table when the actual scenario is revealed to the decision maker, i.e. when it is known with perfect hindsight what the optimal policy should have been. The ideal is to leave no money on the table under any scenario, but this is impossible. Regret is always nonnegative and unless the problem is trivial, no trading policy can achieve zero-regret.

4 SCENARIO OPTIMIZATION

Formally, let us denote by $V_q(y)$ the value in the objective function (4) for the candidate solution y, such that

$$V_q(y)=S_T(y)+\sum_{i\in I}\rho_{iT}\,\bar{P}_{iT}^{q}\,Z_{iT}(y), \tag{17}$$

where $S_T(y)$ and $Z_{iT}(y)$ give the cash and stock of instrument i at the end of the planning horizon due to the trading policy y, respectively. Let also $H_q(y)$ denote the penalty for violating the cash-flow balance equation (6) by the trading policy y.

There are different ways to express the regret, say, $R(y)$ of a trading policy y. Our favorite expression is $R(y)=\|w(V(y)-V(*))\|^d+\|wH(y)\|^d$, where w, $V(y)$, $H(y)$ and $V(*)$ are $|Q|$ vectors, being w_q, $V_q(y)$, $V_q(*)$ and $H_q(y)$, the related elements for scenario q for $q\in Q$, and d is a suitable norm parameter. The zero-regret can be expressed as $R(*)$ where $*$ is the related optimal policy. The regret distribution, say $F(y)$ for y is the distribution of the regret, namely, $F(y)_a=P(R(y)\leqslant a)$ for $a\geqslant 0$, and the zero-regret distribution is $F(*)_a=1$ for $a\geqslant 0$.

A mixed 0–1 program that produces the policy y whose regret distribution most closely approximates the zero-regret distribution for some d is the so-called *scenario immunization* (SI) model (20)–(29), below. (Dembo and King, 1991, study optimal regret distributions for asset allocation in general.)

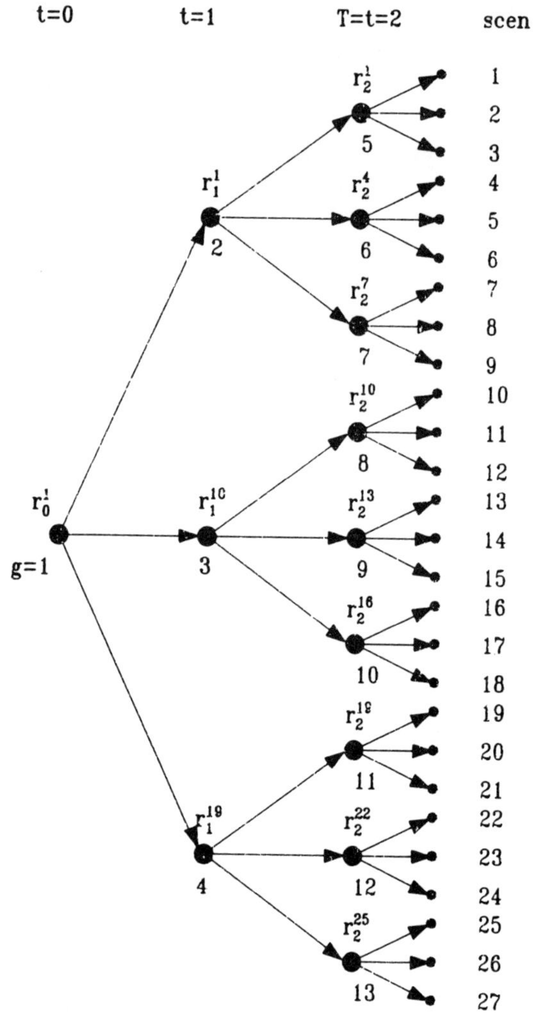

Figure 13.1

Let G_t denote the set of groups of scenarios that start from time period t, such that if two different scenarios, say q and q' are identical up to period t (on the basis of the information available about them at period t, i.e. $r_\tau^q = r_\tau^{q'}$ for $0 \leqslant \tau \leqslant t$), then both scenarios q and q' belong to the same group, say Q_g for $g \in G_t$, $0 \leqslant t \leqslant T$. In Figure 13.1 see that $Q_g = \{10, 11, \ldots, 18\}$ for $g = 3$ and $G_t = \{2, 3, 4\}$ for $t = 1$. Note also that $Q_{g'} \subseteq Q_g$ for $g' \in G_\tau$, $g \in G_t$ and $t < \tau$ as $Q_{10} = \{16, 17, 18\} \subset Q_3$ since $t = 1 < \tau = 2$.

Let $\tilde{P}_{it}^g$ denote the estimated unit market price of instrument i at (the end of) period t under scenario group g for $g \in G_t$, $0 \leqslant t \leqslant T_i$, such that it can be expressed as

$$\tilde{P}_{it}^g = \sum_{q \in Q_g} w_q P_{it}^q. \tag{18}$$

(Note: the rationale under expression (18) is that the market has the same information as the modeler.)

Finally, let the weight of a scenario group, say w_{gt} for $g \in G_t$ be such that

$$\tilde{w}_{gt} = \sum_{q \in Q_g} w_g. \tag{19}$$

The SI model is as follows:

$$\min \sum_{g \in G_T} \tilde{w}_{gT} [S_T + \sum_{i \in I} \rho_{iT} \tilde{P}_{iT}^{-g} Z_{it} - \hat{V}_{gT}(*)]_-^d + \alpha H(s)_- \tag{20}$$

subject to

$$Z_{it} + X_{it}^+ - X_{it}^- = Z_{i,t+1}, \quad i \in I, \quad 0 \leqslant t < T, \tag{21}$$

$$-S_{t-1} + \sum_{i \in I} [-C_{it} Z_{it} + \rho_{it} \tilde{P}_{it}^{+g} X_{it}^+ - \rho_{it} \tilde{P}_{it}^{-g} X_{it}^-] + S_t - s_t^g = D_t, \quad g \in G_t, \ 0 \leqslant t < T, \tag{22}$$

$$S_{T-1} + \sum_{i \in I} C_{iT} Z_{iT} = S_T, \tag{23}$$

$$m_{jt} \delta_{jt} \leqslant \sum_{i \in I_j} F_i Z_{it} \leqslant M_{jt} \delta_{jt}, \quad j \in J, \quad 0 < t \leqslant T, \tag{24}$$

$$\sum_{j \in J} \delta_{jt} \leqslant N_t, \quad 0 < t \leqslant T, \tag{25}$$

$$0 \leqslant X_{it}^+ \leqslant \overline{X}_{it}^+, \quad 0 \leqslant X_{it}^- \leqslant \overline{X}_{it}^-, \quad i \in I, \quad 0 \leqslant t < T, \tag{26}$$

$$\delta_{jt} \in \{0; 1\}, \quad j \in J, \quad 0 < t \leqslant T, \tag{27}$$

$$0 \leqslant Z_{it}, \quad i \in I, \quad K_i \leqslant t \leqslant T_i, \tag{28}$$

$$0 \leqslant S_t, \quad 0 \leqslant t \leqslant T. \tag{29}$$

where d and α are positive parameters, $P^{+g} \equiv P^g + \alpha$, $P^{-g} \equiv P^g - \alpha$, $s_0^g = 0$, and s_t^g $\forall g,t$ is an unrestricted variable that takes the violation of the cash-flow balance equation for period t under scenario group g. $H(s)_-$ is the penalty function of $\{(S_t^g)_- \ \forall g,t\}$ (see below), and $(r)_-$ denotes the lower partial value of variable r, such that

$$(r)_- = \begin{cases} 0 & \text{for } r \geqslant 0, \\ |r| & \text{otherwise.} \end{cases} \tag{30}$$

Note that the positive value of s_t^g (i.e. the cash surplus in period t under scenario group q) is not penalized; obviously, the consequences of $s_t^g > 0$ are not as bad as the consequences of cash expediting (i.e. $s_t^g < 0$).

On the other hand, $\hat{V}_{qT}(*)$ gives the optimality tolerance (see Jensen and King, 1991) such that

$$\hat{V}_{gT}(*) = \tilde{V}_{gT}(*) - \epsilon \quad \text{for} \quad \epsilon \geqslant 0, \tag{31}$$

where $\tilde{V}_{gT}(*)$ takes the terminal wealth of the optimal trading policy under scenario group g for $g \in G_T$ at the end of the planning horizon, provided that the market prices the financial instruments as in the above model. So, $\tilde{V}_{gT}(*)$ can be obtained as follows:

$$\tilde{V}_{gT}(*) = \max S_T + \sum_{i \in I} \rho_{it} \tilde{P}_{iT}^{-g} Z_{iT}, \tag{32}$$

subject to (5)–(13), where P_{it}^{+} and P_{it}^{-} in (6) are replaced by $\tilde{P}_{it}^{+g'}$ and $\tilde{P}_{it}^{-g'}$, respectively, for $g' \in G_t$ the scenario group for $0 \leqslant t < T$, such that $Q_g \subseteq Q_{g'}$.

The objective function (20) is a tracking scenario function (see Dembo, 1991). It minimizes the regret of wrong trading decisions. The regret is defined as the expected lower partial difference between the terminal wealth of the model's trading policy (X^+, X^-) and the terminal wealth of the optimal trading policy under any scenario group at the end of the planning horizon. By definition, the difference for each scenario group is nonpositive (usually, negative) for $H(s)_- = 0$; in any case, we are more interested in the lower partial value. The main difference between the programs (4)–(13) and (20)–(29) is the set of constraints for the cash-flow balance equations. The constraints in the SI model allow one to consider the residual value (s_t^g) for each period and scenario group; the number of constraints is

$$\sum_{0 \leqslant t < T} \| G_t \|.$$

The objective function (20) is a composite function of a penalization for cash-flow balance equations' lower partial infeasibility, and the expected lower partial deviation from the optimal terminal wealth under any scenario group at the end of the planning horizon.

Mulvey *et al.* (1991) suggest several forms in which to express a penalty function H such as ours. They may be expressed as follows:

$$H(\xi) \; : \; \max_{0<t<T}\{\xi_t\} \qquad , \tag{33}$$
$$\; : \; \Sigma_{0<t<T}\, \xi_t^2$$

where

$$\xi_t = \sum_{g \in G_t} \tilde{w}_{gt} (s_t^g)_- . \tag{34}$$

A usual way to model $(s_t^g)_-$ is to replace s_t^g by $s_t^{+g} - s_t^{-g}$ in (22), where $0 \leqslant s_t^{+g}$, s_t^{-g} such that ξ_t (34) can be expressed as $\xi_t = \Sigma_{g \in G_t} \tilde{w}_{gt} s_t^{-g}$.

The SI model (20)–(29) allows one to consider the uncertainty in the interest rate path along the planning horizon. The dimensions are bigger than the dimensions of (4)–(13). Depending upon the norm parameter d, the function (20) may be nonlinear. However, given an appropriate meaning for d, we can express (20) in a linear form. Specifically, a hierarchical two-step approach can be proposed. It is as follows.

1. Minimize the weighted maximum regret:

$$\min \max_{g\in G_T}\{\tilde{w}_{gT}(\tilde{V}_{gT}-\hat{V}_{gT}(*))_-\}+\alpha \max_{0<t<T}\{\xi_t\}|(21)\text{–}(29) \tag{35}$$

where ξ_t is given by (34) and

$$\tilde{V}_{gT}\equiv S_T+\sum_{i\in I}\rho_{iT}\tilde{P}_{iT}^{-g}Z_{iT}. \tag{36}$$

Let $\overline{V}$ and $\overline{\xi}$ be such that $\exists g\in G_T|\overline{V}=\tilde{w}_{gT}(\tilde{V}_{gT}\hat{V}_{gT}(*))_-$ and $\exists t|\overline{\xi}=\xi_t$, $0<t<T$.

2. Minimize the total weighted regret within the bounds $\overline{V}$ and $\overline{\xi}$:

$$\min \sum_{g\in G_T} \tilde{w}_{gT}(\tilde{V}_{gT}-\hat{V}_{gT}(*))_- +\alpha \sum_{0<t<T} \xi_t|(21)\text{–}(29), \tag{37}$$

subject to

$$\tilde{w}_{gT}(\tilde{V}_{gT}-\hat{V}_{gT}(*))_-\leqslant\overline{V},\quad g\in G_T, \tag{38}$$

$$\xi_t\leqslant\overline{\xi},\quad 0<t<T. \tag{39}$$

As an alternative objective function to (20), we may have

$$\min \sum_{g\in G_T} \tilde{w}_{gT}(\tilde{V}_{gT}-\tilde{V}_{gT}(*))^d+\alpha H(s), \tag{40}$$

where

$$H(s):\frac{1}{T-1}\sum_{0<t<T}\left(\xi_t-\frac{1}{T-1}\sum_{0<\tau<T}\xi_\tau\right)^2, \tag{41}$$

See in Golub *et al.* (1991) a stochastic programming model whose objective function to maximize is the expected terminal wealth.

5 IMPLEMENTABLE POLICIES

We should point out that the SI model (20)–(29) and variations do anticipate trading decisions that may not be needed at time period $t=0$.

Frequently, the decisions related to period $t=0$ are the only decisions to make. The rationale behind this policy is that (i) the data for the model may have changed prior to the next trading session (due to new offers and actions in the market, change in the exogenous cash availability, news that influences the interest rate and then the scenario's structure, etc.); and (ii) the modeler still has time to produce and analyze the new trading policy. So, the model can be reoptimized at the end of each period in a rolling-horizon mode. (See that after implementing the spot trading decisions X_{i0}^{+} and X_{i0}^{-}, we have available automatically the cash S_0 at the end of period 0 and the instruments' stock Z_{i1} at the beginning of period $t=1$.) See also that, in this case, the information about future uncertainty is only taken into account for deciding a better spot trading. This type of approach is termed *full recourse*. The SI model has been designed with a *simple recourse* philosophy in mind. Any intermediate situation is termed *partial recourse*, see Rockafellar and Wets (1987), Wets (1989), and Escudero *et al.* (1993), among others.

Rockafellar and Wets (1987) state the implementable or nonanticipativity principle for full-recourse modeling. Simply stated it is as follows. 'If two different scenarios, say q and q', are identical up to time period τ on the basis of the information available about them at time period τ, then the control variables X_{it}^{+} and X_{it}^{-} (and, then, the related state Z-variables) must also be identical for $0 \leqslant t \leqslant \tau$. This condition guarantees that the solutions obtained by the model do not depend (at time period τ) on information that is not yet available. It is best visualized as follows. Each node in Figure 13.1 represents a point in time where a decision can be made. Once a decision is made, one of several contingencies can happen (in this example, the contingencies are three) and this information is available at the beginning of the next time period. Each root-to-leaf path in Figure 13.1 represents one scenario and it is one realization of the interest rate over time.

Consider time period 1 for its three scenario groups shown in Figure 13.1. At this stage, a decision has to be made as to how much to trade of financial instrument i at the end of period 1. This is naturally a function of the scenario group that we may follow and, then, according to the implementable principle, our notation for X^{+}, X^{-}, Z and S is as follows: X_{it}^{+g}, X_{it}^{-g}, Z_{it}^{g}, S_{it}^{g} give the related variables, respectively, for financial instrument i, scenario group g and time period t for $i \in I$, $g \in G_t$, and $0 \leqslant t \leqslant T$. See Figure 13.2.

Let $\tilde{V}_{gT}$ for $g \in G_T$ be such that

$$\tilde{V}_{gT} \equiv S_T^g + \sum_{i \in I} \rho_{iT} \tilde{P}_{iT}^g Z_{iT}^g . \tag{42}$$

It denotes the terminal wealth under scenario group g for $g \in G_T$.

Let $\tilde{V}_{gT}(*)$ take the terminal wealth of the optimal trading policy under scenario group g for $g \in G_T$, provided that the market prices the financial instruments, such that

$$\tilde{P}_{iT}^{g'} = \sum_{q \in Q_{g'}} w_q P_{it}^q, \quad i \in I,\ g' \in G_t | Q_g \subseteq Q_{g'}, \quad 0 \leqslant t \leqslant T. \tag{43}$$

Our *full-recourse* (FR) program to maximize the expected terminal wealth is as follows:

$$\max \sum_{g \in g_T} \tilde{w}_{gT} \tilde{V}_{gt} | (46)\text{–}(54). \tag{44}$$

Our FR program with scenario immunization can be expressed as follows:

$$\min \sum_{g \in G_T} \tilde{w}_{gT} (\tilde{V}_{gT} - \tilde{V}_{gT}(*))^d \tag{45}$$

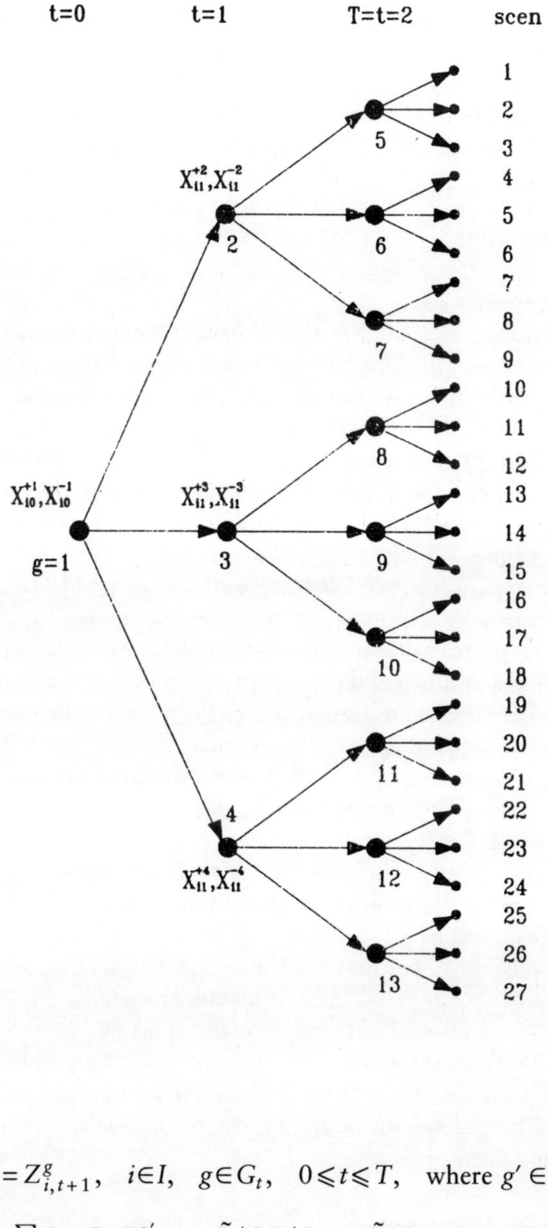

Figure 13.2

subject to

$$Z_{it}^{g'} + X_{it}^{+g} - X_{it}^{-g} = Z_{i,t+1}^{g}, \quad i \in I, \quad g \in G_t, \quad 0 \leqslant t \leqslant T, \quad \text{where } g' \in G_{t-1}, \tag{46}$$

$$\begin{aligned} -S_{t-1}^{g'} + \sum_{i \in I} [-C_{it} Z_{it}^{g'} + \rho_{it} \tilde{P}_{it}^{+g} X_{it}^{+g} - \rho_{it} \tilde{P}_{it}^{-g} X_{it}^{-g}] + S_t^g = D_t, \\ g \in G_t, \quad 0 \leqslant t < T, \quad \text{where } g' \in G_{t-1}, \end{aligned} \tag{47}$$

$$-S_{T-1}^{g'} + \sum_{i \in I} C_{it} Z_{iT}^{g'} = S_t^g, \quad g \in G_T, \quad \text{where } g' \in G_{T-1}, \tag{48}$$

$$m_{jt} \delta_{jt}^g \leqslant \sum_{i \in I_j} F_i Z_{it}^g \leqslant M_{jt} \delta_{jt}^g, \quad j \in J, \quad g \in G_t, \quad 0 < t \leqslant T, \tag{49}$$

$$\sum_{j\in J}\delta_{it}^{g}\leqslant N_t,\quad g\in G_t,\quad 0<t\leqslant T, \tag{50}$$

$$0\leqslant X_{it}^{+g}\leqslant \overline{X}_{it}^{+},\quad 0\leqslant X_{it}^{-g}\leqslant \overline{X}_{it}^{-},\quad i\in I,\quad g\in G_t,\quad 0\leqslant t<T, \tag{51}$$

$$\delta_{jt}^{g}\in\{0;\ 1\},\quad j\in J,\quad g\in G_t,\quad 0<t\leqslant T, \tag{52}$$

$$0\leqslant Z_{it}^{g},\quad i\in I,\quad g\in G_t,\quad K_i\leqslant t\leqslant T_i, \tag{53}$$

$$0\leqslant S_t^{g},\quad g\in G_t,\quad 0\leqslant t\leqslant T. \tag{54}$$

The meaning of the objective function and constraints above is very similar to the meaning of the SI model (20)–(29). Note that the penalty function H is not required in (45). See that the dimensions of the FR model are much bigger than the dimensions of the SI model.

The FR model (45)–(54) can also be expressed as a mixed 0–1 linear program by using the hierarchical two-step approach (35)–(39). Anyway, we may exploit its nice structure. Two approaches can be used for expressing the implementable principle for the Z- and S-variables. Approach 1 is based on a *compact* representation, see Escudero and Kamesam (1992), where the variables are used as formally expressed in (46)–(48). It allows for a nice crash optimization and it is more suitable for the simplex methodology for pure linear programs. Approach 2 is based on a *splitting variable* representation, see Lustig *et al.* (1991), and Escudero and Kamesam (1992), where the variables are split such that new sets of constraints are introduced as $Z_{it}^{g'}=Z_{it}^{g}$ and $S_{t-1}^{g'}=S_{t-1}^{g}$ where $g'\in G_{t-1}$ and $g\in G_t$ and, then, each constraint can only have variables that belong to the same scenario group, except for the so-called coupling constraints just given. It is more suitable for the interior-point methodology (it is larger but also preserves some sparsity) and, as in our case, for models with 0–1 variables. Anyway, the approach allows for decomposition techniques such as the progressive hedging algorithm introduced by Rockafellar and Wets (1987) and Wets (1989), see computational results in Mulvey and Vladimirou (1989) among others, and variations of Benders and Dantzig–Wolfe decompositions, see Alvarez *et al.* (1993).

An additional advantage of the splitting-variable representation (specially for 0–1 variables) is the possibility of using Lagrangian decomposition techniques; see Jörnsten *et al.* (1985) and Guignard and Kim (1987). Bjornestad *et al.* (1992) describe a Lagrangian decomposition approach for an SP model with implementable policies to obtain.

6 DUALIZING THE FULL-RECOURSE MODEL

To simplify the notation let us group the time periods in three stages, such that stage 1 is included by time period $t=0$, stage 2 is time period 1, and stage 3 is included by all periods $\{t\}$ for $1<t\leqslant T$. We also assume that the interest rate for time period 0 is known at period $t=0$, and three outlooks are considered for each of stages 2 and 3.

The following notation is needed for the (compact) representation of the deterministic program for the three stages.

Y^0, set of X-variables for stage 1, and S-variables for stage 1 that are not linking variables between stages 1 and 2 (i.e. variables for stage 1 with no nonzero elements in constraints for periods in stage 2).
Z^0, set of Z-variables for stage 1 that are not linking variables between stages 1 and 2.
S_{12} and Z_{12}, sets of linking S-variables and Z-variables between stages 1 and 2, respectively.
δ^0, sets of δ-variables for stage 1.
Y_1 and Z_1, sets of X-, S- and Z-variables for stage 2 that are not linking variables between stages 1 and 2, not between stages 2 and 3, respectively.
S_{23} and Z_{23}, sets of linking S-variables and Z-variables between stages 2 and 3, respectively.
δ^1, set of δ-variables for stage 2.
Y_2 and Z_2, sets of X-, S- and Z-variables for stage 3 that are not linking variables between stages 2 and 3, respectively.
δ_2, set of δ-variables for stage 3.
A^0, B^0, C^0, and D^0, appropriate matrices for variables Y^0, Z^0, S_{12}, and Z_{12} in constraints (5)–(7), respectively.
F^0 and F_{12}, appropriate 0–1 matrices for variables Z^0 and Z_{12} in constraints (8), respectively.
M^0, appropriate matrix for variables δ^0 in constraints (8). Note that there is only one nonzero element per row, and there are two nonzero (one negative and the other positive) elements per column.
I^0, appropriate unit matrix for variables δ^0 in constraints (9).
C^1_{12}, D^1_{12}, A_1, B^1, C^1_{23}, and D^1_{23}, appropriate matrices for variables S_{12}, Z_{12}, Y_1, Z_1, S_{23}, and Z_{23} in constraints (5)–(7), respectively.
F^1, F_{23}, and M^1, appropriate matrices for variables Z_1, Z_{23}, and δ_1 in constraints (8), respectively.
I^1, appropriate unit matrix for variables δ_1 in constraints (9).
C^2_{23}, D^2_{23}, A_2 and B^2, appropriate matrices for variables S_{23}, Z_{23}, Y_2, and Z_2 in constraints (5)–(7), respectively.
F^1 and M^2, appropriate matrices for variables Z_2 and δ_2 in constraints (8), respectively.
I^2, appropriate unit matrix for variables δ_2 in constraints (9).
b^0, b^1, and b^2, appropriate right-hand-side vectors in constraints (5)–(7) for stages 1, 2, and 3, respectively.
N^0, N^1, and N^2, appropriate right-hand-side vectors in constraints (9) for stages 1, 2, and 3, respectively.

Letting $f(Y, Z)$ denote the objective function to optimize in (4), the deterministic mixed 0–1 program (4)–(13) can be expressed as follows:

$$
\begin{array}{l}
D\text{: } \min f(Y, Z) \\
\text{subject to}
\end{array}
$$

$$
\begin{array}{lr}
A^0 Y^0 + B^0 Z^0 + C_{12}^0 S_{12} + D_{12}^0 Z_{12} & \leqslant b^0, \\
F^0 Z^0 + F_{12} Z_{12} - M^0 \delta^0 & \leqslant 0, \\
I^0 \delta^0 & \leqslant N_0, \\
C_{12}^1 S_{12} + D_{12}^1 Z_{12} + A_1 Y_1 + B^1 Z_1 + C_{23}^1 S_{23} + D_{23}^1 Z_{23} & \leqslant b^1, \\
F^1 Z_1 + F_{23} Z_{23} - M^1 \delta_1 & \leqslant 0, \\
I^1 \delta_1 & \leqslant N_1, \\
C_{23}^2 S_{23} + D_{23}^2 Z_{23} + A_2 Y_2 + B^2 Z_2 & \leqslant b^2, \\
F^2 Z_2 - M^2 \delta_2 & \leqslant 0, \\
I^2 \delta_2 & \leqslant N_2, \\
Y^0, \quad Z^0, \quad S_{12}, \quad Z_{12}, \quad Y_1, \quad Z_1 \quad S_{23}, \quad Z_{23}, \quad Y_2, \quad Z_2, & \geqslant 0, \\
\delta_0, \quad \delta_1, \quad \delta_2 & \in \{0;\ 1\}
\end{array}
$$

The full-recourse (FR) model related to the deterministic model given above, requires the additional notation of Table 13.1

Table 13.1

Deterministic model	Full-recourse model
S_{12}	S_{12}^0, S_{12}^k
Z_{12}	Z_{12}^0; Z_{12}^k
A_1	A^k
Y_1	Y^k
Z_1	Z^k
S_{23}	S_{23}^{k0}, S_{23}^{kl}
Z_{23}	Z_{23}^{k0}, Z_{23}^{kl}
δ_1	δ^k
A_2	A^{kl}
Y_2	Y^{kl}
Z_2	Z^{kl}
δ_2	δ^{kl}

Additionally, let the unit matrices I_{12}^S, I_{12}^Z, I_{23}^S and I_{23}^Z have the appropriate dimensions for the coupling constraints to be given below; see (55)–(58). Finally, let $F(Y, Z)$ denote the objective function (45). It can be expressed as (35) or (37).

The full-recourse mixed 0–1 program (45)–(54) is as follows:

$$I_{12}^S S_{12}^0 - I_{12}^S S_{12}^k = 0, \quad (55)$$

$$I_{12}^Z Z_{12}^0 - I_{12}^Z Z_{12}^Z = 0, \quad (56)$$

$$I_{23}^S S_{23}^{k0} - I_{23}^S S_{23}^{kl} = 0, \quad (57)$$

$$I_{23}^Z Z_{23}^{k0} - I_{23}^Z Z_{23}^{kl} = 0. \quad (58)$$

The coupling constraints (55)–(58) cause the major obstacle for decomposing the FR model. An *augmented Lagrangian* methodology overcomes this difficulty by dualizing them. The resulting problem is;

$$\max_{\lambda,\mu} \mathrm{FR}_p(\lambda, \mu), \quad (59)$$

where

$$\begin{aligned}\mathrm{FR}_p(\lambda, \mu) = \min F(Y, Z) \quad & + \lambda^0 S_{12}^0 - \Sigma_{k=1,2,3} \lambda^k S_{12}^k \\ & + \mu^0 Z_{12}^0 - \Sigma_{k=1,2,3} \mu^k Z_{12}^k \\ & + \Sigma_{k=1,2,3} \lambda^{k0} S_{23}^{k0} - \Sigma_{k=1,2,3} \lambda^{kl} S_{23}^{kl} \\ & + \Sigma_{k=1,2,3} \mu^{k0} Z_{23}^{k0} - \Sigma_{k,l=1,2,3} \mu^{kl} Z_{23}^{kl} \\ + (p/2)(\Sigma_{k=1,2,3}(S_{12}^0 - S_{12}^k)^2 & + \Sigma_{k,l=1,2,3}(S_{23}^{k0} - S_{23}^{kl})^2) \\ + (p/2)(\Sigma_{k,l=1,2,3}(Z_{12}^0 - Z_{12}^k)^2 & + \Sigma_{k,l=1,2,3}(Z_{23}^{k0} - Z_{23}^{kl})^2)\end{aligned} \quad (60)$$

FR: $\min f(Y, Z)$

subject to

$$
\begin{array}{llll}
A^0Y^0 + B^0Z^0 + C^0_{12}S^0_{12} + D^0_{12}Z^0_{12} & & \leqslant b^0, & \\
F^0Z^0 + F_{12}Z^0_{12} - M^0\delta^0 & & \leqslant 0, & \\
I^0\delta^0 & & \leqslant N^0, & \\
I^s_{12}S^0_{12} - I^s_{12}S^k_{12} & & = 0 & \forall k, \\
I^s_{12}Z^0_{12} - I^z_{12}Z^k_{12} & & = 0 & \forall k, \\
C^1_{12}S^k_{12} + D^1_{12}Z^k_{12} + A^kY^k + B^1Z^k + C^1_{23}S^{K0}_{23} + D^1_{23}Z^{K0}_{23} & & \leqslant b^1 & \forall k, \\
F^1Z^k + F_{23}Z^{K0}_{23} - M^1\delta^k & & \leqslant 0 & \forall k, \\
I^1\delta^k & & \leqslant N^1 & \forall k, \\
I^S_{23}S^{K0}_{23} - I^S_{23}S^{kl}_{23} & & = 0 & \forall k,l. \\
I^z_{23}Z^{k0}_{23} - I^z_{23}Z^{kl}_{23} & & = 0 & \forall k,l. \\
C^2_{23}S^{kl}_{23} + D^2_{23}Z^{kl}_{23} + A^{kl}Y^{kl} + B^2Z^{kl} & & \leqslant b^2 & \forall k,l. \\
F^2Z^{kl} - M^2\delta^{kl} & & \leqslant 0 & \forall k,l. \\
I^2\delta^{kl} & & \leqslant N^2 & \forall k,l. \\
Y^0,\ Z^0,\ S^0_{12},\ Z^0_{12},\ S^k_{12},\ Z^k_{12},\ Y^k,\ Z^k\ S^{k0}_{23},\ Z^{k0}_{23},\ S^{kl},\ Z^{kl}_{23},\ Y^{kl},\ Z^{kl}, & & \geqslant 0 & \forall k,l. \\
\delta^0,\ \delta^k\ \delta^{kl} & & \in \{0;\ 1\} & \forall k,l.
\end{array}
$$

subject to the (independent) constraint systems (67), (68), and (69), and where

$$\lambda = (\lambda^k, \lambda^{k,l}, \quad \forall k, l = 1,2,3), \tag{61}$$

$$\mu = (\mu^k, \lambda^{k,l}, \quad \forall k, l = 1,2,3) \tag{62}$$

are the vectors of the dual variables for the constraints (55)–(56) and (57)–(58), respectively, $p > 0$ is a penalty parameter, and

$$\lambda^0 = \sum_{k=1,2,3} \lambda^k, \tag{63}$$

$$\lambda^{k0} = \sum_{l=1,2,3} \lambda^{kl}, \quad k = 1,2,3, \tag{64}$$

$$\mu^0 = \sum_{k=1,2,3} \mu^k, \tag{65}$$

$$\mu^{k0} = \sum_{l=1,2,3} \mu^{kl}, \quad k = 1,2,3. \tag{66}$$

subject to (69).

An iterative procedure may solve problem (59) by iteratively solving successive problems (60) for the set of parameters (p, λ, μ). At each (so-called major) iteration the parameters are updated in the usual way; i.e. p is appropriately reduced such that $p \to 0$, and the multipliers λ and μ are updated based on the infeasibility of the constraints (55)–(58), see Bertsekas (1982).

Note that a given problem (60) has independent sets of constraints, but a quasiseparable QP function. Let $\overline{R}$ denote a given value for vector R. Problem (60) can be solved by iteratively solving successive sets of separable QP mixed 0–1 programs (67), (68), and (69). For this purpose, at each (so-called minor) iteration the solution $(\overline{S}, \overline{Z})$ from the previous iteration is used.

$$L^0 = \min \lambda^0 S_{12}^0 + \mu^0 Z_{12}^0 + (p/2) \sum_{k=1,2,3} ((S_{12}^0 - \overline{S}_{12}^k)^2 + (Z_{12}^0 - \overline{Z}_{12}^k)^2), \tag{67}$$

subject to (70).

$$\begin{aligned} L^k = \min & -(\lambda^k S_{12}^k + \mu^k Z_{12}^k) + (\lambda^{k0} S_{23}^{k0} + \mu^{k0} Z_{23}^{k0}) \\ & + (p/2)((\overline{S}_{12}^0 - S_{12}^k)^2 + (\overline{Z}_{12}^0 - Z_{12}^k)^2) \\ & + (p/2)\Sigma_{l=1,2,3}((S_{23}^{k0} - \overline{S}_{23}^{kl})^2 + (Z_{23}^{k0} - \overline{Z}_{23}^{kl})^2), \end{aligned} \tag{68}$$

subject to (71).

$$\begin{aligned} L^{kl} = \min & \ F(Y^{kl}, Z^{kl}) - (\lambda^{kl} S_{23}^{kl} + \mu^{kl} Z_{23}^{kl}) \\ & + (p/2)((\overline{S}_{23}^{k0} - S_{23}^{kl})^2 + (\overline{Z}_{23}^{k0} - Z_{23}^{kl})^2), \end{aligned} \tag{69}$$

subject to (72)

$$
\begin{aligned}
&A^0Y^0+B^0Z^0+C^0S^0_{12}+D^0_{12}Z^0_{12} && \leqslant b^0,\\
&F^0Z^0 \qquad\qquad F_{12}Z^0_{12}-M^0\delta^0 && \leqslant 0,\\
& I^0\delta^0 && \leqslant N^0,\\
&Y^0,\quad Z^0,\quad S^0_{12},\quad Z^0_{12} && \geqslant 0,\\
& && \delta^0\in\{0;\ 1\}.
\end{aligned} \tag{70}
$$

$$
\begin{aligned}
&C^1_{12}S^k_{12}+D^1_{12}Z^k_{12}+A^kY^k+B^1Z^k+C^1_{23}S^{k0}_{23}+D^1_{23}Z^{k0}_{23} && \leqslant b^1,\\
&F^1Z^k \qquad + F_{23}Z^{k0}-M^1\delta_k && \leqslant 0,\\
&I^1\delta^k && \leqslant N^1 \quad \forall k,\\
&S^k_{12},\quad Z^k_{12},\quad Y^k,\quad Z^k,\quad S^{k0}_{23};\quad Z^{k0}_{23} && \geqslant 0,\\
& && \delta_k\in\{0;\ 1\}.
\end{aligned} \tag{71}
$$

$$
\begin{aligned}
&C^2_{23}S^{kl}_{23}+D^2_{23}Z^{kl}_{23}+A^{kl}Y^{kl}+B^2Z^{kl} && \leqslant b^2,\\
&F^2Z^{kl}-M^2\delta^{kl} && \leqslant 0,\\
&I^2\delta^{kl} && \leqslant N^2 \quad \forall k,l,\\
&S^{kl}_{23},\quad Z^{kl}_{23},\quad Y^{kl},\quad Z^{kl} && \geqslant 0,\\
& && \delta^{kl}\in\{0;\ 1\}.
\end{aligned} \tag{72}
$$

Note that the constraint systems (70), (71), and (72) have a similar structure. They are very sparse linear systems with two types of logical constraints. The Z-variables are semicontinuous variables and the number of variables with nonzero value for disjoint subsets of Z-variables is upper-bounded. In any case, given the proposed decomposition, the number of independent programs to be solved at each iteration is $1+|G_1|(1+|G_2|)$, where $|G_1|$ and $|G_2|$ are the numbers of interest-rate outlooks for stages 2 and 3, respectively (see Figure 13.1).

Along the framework described in Mulvey and Ruszczynski (1992), Ruszczynski (1992), and Escudero (1993), an alternative approach for obtaining separable QP programs consists of approximating the quadratic term, say $(a-b)^2$ such that the approximation is QP separable with error $O(a-b)^2$. Note that $ab=\bar{a}b+a\bar{b}-\bar{a}\bar{b}$, where (a, b) is a point in the neighborhood of some reference point $(\bar{a}, \bar{b})$, see Stephanopoulos and Westerberg (1975). Then, we approximate

$$(a-b)^2 \approx a^2+b^2-2\bar{a}b-2a\bar{b}+\bar{a}\bar{b}. \tag{73}$$

Anyway, even a separable QP mixed 0–1 program can be difficult to solve for large-scale cases. We are experimenting with a further approximation based on the linearization of a^2 and b^2. The typical expression has the form

$$(a-b)^2 \approx -(\bar{a}-\bar{b})^2+2(\bar{a}-\bar{b})a+2(\bar{b}-\bar{a})b, \tag{74}$$

The resulting function to optimize is a so-called *reinforced Lagrangian* function.

For problems without logical constraints (i.e. continuous FR models), the interior-point methodology is well suited to solving separable QP; see Carpenter *et al.* (1993). This is another direction for further research. Note that the integrality constraint of the 0–1 variables related to the last stage can be relaxed without too much loss of accuracy, given the uncertain nature of the data for the last periods of the planning horizon.

7 CONCLUSIONS

A model for trading financial instruments with uncertainty in interest rates along a planning horizon is proposed. The uncertainty in the interest rate is taken into account by considering a set of scenarios. Both types of multi-stage approach, namely simple recourse and full recourse are analyzed. We conclude that full recourse has more theoretical advantages than simple recourse for this type of problem. The goal then is to minimize the expected (partial lower) differencc between the terminal wealth of the proposed solution and the optimal terminal wealth under any scenario, such that the nonanticipativity principle is preserved. The optimization is performed by using a two-step procedure as follows: first, the greatest terminal wealth deviation is minimized and, then, the total expected deviation from the optimal terminal wealth is minimized subject to the results of the first step. The resulting LP models are very large, but the splitting-variable constraints can be dualized. An *augmented Lagrangian* quasiseparable QP function subject to independent sets of constraints is obtained. By linearizing cross-products, the function can be approximated by a separable QP function and, then, independent QP mixed 0–1 problems can be optimized. As an alternative for large-scale mixed 0–1 programs, the QP function can itself be linearized and, then, it is approximated by a so-called *reinforced Lagrangian* function. Given the uncertain nature of the data for the last periods of the planning horizon, the integrality constraint of the 0–1 variables related to the last stage can be relaxed without too much loss of accuracy. A careful implementation is under way to assess the approach presented in this chapter.

REFERENCES

Alvarez, M., Cuevas, C. M., Escudero, L. F., de la Fuente, J. L., García, C., and Prieto, J. L. (1993) Linear networks with uncertainty. An application field for parallel computing in optimization, *Netflow 93, Pisa, Italy*.

Bertsekas, D. P. (1982) *Constrained Optimization and Lagrange Multipliers*, Academic Press, New York.

Bjornestad, S., Hallefjord, A., and Jörnsten, K. (1992) Discrete optimization under uncertainty: The scenario and policy aggregation technique, *European Journal of Operational Research*.

Carpenter, T. J., Lustig, I. J. Mulvey, J. M., and Shanno, D. F. (1993) Separable quadratic programming via a primal–dual interior method and its use in a sequential procedure, *ORSA Journal on Computing*, 5, 182–191.

Dembo, R. S. (1991) Scenario optimization, *Annals of Operations Research*, 30, 63–80.

Dembo, R. S., and King, A. J. (1991) *Tracking Models and the Optimal Regret Distribution in Asset Allocation*, IBM T. J. Watson Research Center, RC-17156, Yorktown Heights, NY.

Escudero, L. F. (1993) *Robust MBS Portfolios*, technical report, Mathematical Sciences School, Universidad Complutense de Madrid, Spain.
Escudero, L. F., and Kamesan, P. V. (1992) *On Solving Stochastic Models via Scenario Analysis*, technical report, IBM Research T. J. Watson Research Center, Yorktown Heights, NY.
Escudero, L. F., Kamesan, P. V., King, A., and Wets, R. J.-B. (1993) Production planning via scenario modelling, *Annals of Operations Research*, **43**.
Golub, V., Holmer, M., MacKendall, R., Pohlman, L., and Zenios, S.A. (1991) *A Stochastic Programming Model for Money Management*, Report 93-01-01, Wharton School, University of Pennsylvania, PA.
Guignard, M., and Kim, S. (1987) Lagrangean decomposition: A model yielding stronger Lagrangean bounds, *Mathematical Programming*, **39**, 215–228.
Jensen, D., and King, A. (1991) *Linear-Quadratic Efficient Frontiers for Portfolio Optimization*, IBM T. J. Watson Research Center, RC-17156, Yorktown Heights, NY.
Jörnsten, K. O., Näsberg, M., and Sneds, P. A. (1985) *Variable Splitting—A New Lagrangean Relaxation Approach to some Mathematical Programming Models*, Report LiTH-MATR-85-04 Linkoping Institute of Technology, Sweden.
Kusy, M. I., and Ziemba, W. T. (1986) A bank asset and liability management model, *Operations Research*, **34**, 356–376.
Lustig, I. J., Mulvey, J. M., and Carpenter, T. J. (1991) Formulating stochastic programs for interior point methods, *Operations Research*, **39**, 757–769.
Mulvey, J. M., and Ruszczynski, A. (1992) A diagonal quadratic approximation model for investment planning, *Operations Research Letters*, **12**, 205–215.
Mulvey, J. M. and Vladimirou, H. (1989) Stochastic network optimization models for investment planning, *Annals of Operations Research*, **20**, 187–217.
Mulvey, J. M., Vanderbei, R. J., and Zenios, S. A. (1991) *Robust Optimization of Large Scale Systems: General Modelling Framework and Computations*, Report 91-06-04, Wharton School, University of Pennsylvania, Philadelphia, PA.
Rockafellar, R. T., and Wets, R. J.-B. (1987) *Scenario and Policy Aggregation in Optimization under Uncertainty*, Working Paper WP-87-119, IIASA, Laxenburg, Austria. See also *Mathematics of Operations Research*, **16** (1991), 119–147.
Ruszezynsky, A. (1993) *An Interior Point Method in Stochastic Programming*, Working Paper WP-93-8, IIASA, Laxenburg, Austria.
Stephanopoulos, G., and Westerberg, W. (1975) The use of Hestene's method of multipliers to resolve dual gaps in engineering system optimization, *Journal of Optimization Theory and Applications*, **15**, 285–309.
Wets, R. J.-B. (1989) The aggregation principle in scenario analysis and stochastic optimization, in: S. W. Wallace (Ed.), *Algorithms and Model Formulations in Mathematical Programming*, Springer pp. 92–113.
Worzel, K. J., and Zenios, S. A. (1992) *Tracking a Mortgage Index: An Optimization Approach*, Technical Report 92-08-91, Wharton School, University of Pennsylvania, Philadelphia, PA.
Zenios, S. A. (1991) Massively parallel computations for financial planning under uncertainty, in: J. P. Mesirov (Ed.), *Very Large Scale Computation in the 21st Century*, SIAM, Philadelphia, pp. 273–294.
Zenios, S. A. (1992) *Asset/Liability Management under Uncertainty: The Case of Mortgage-Backed Securities*, Report 92-08-05, Whaton School, University of Pennsylvania, Philadelphia, PA.

S. Ahn and **M. Guignard-Spielberg**
Operations and Information Management
The Wharton School of the University of Pennsylvania
1300 Steinberg Hall—Dietrich Hall
Philadelphia
PA 19104-6366
USA

L. F. Escudero
Departamento de Estadistiea e I.O.
Universidad Complutense de Madrid
Ciudad Universitaria
28040 Madrid
SPAIN

14 Integrated Planning and Scheduling Systems for the Refining Industry

Ronald E. Coxhead
Bonner & Moore Associates GmbH, Wiesbaden, Germany

ABSTRACT

In many ways the refining industry was the investor and leader in the early development of linear programming in the industrial sector. Many of the applications in long- and medium-term planning, from raw-material selection and refinery-operations optimization through to final product blending were modeled and formed a basis for the development of what are now traditional linear programming or successive LP techniques. A greater focus on crude-oil trading rather than long-term stable supply contracts coupled with narrower refining margins now requires short-term data collection, analysis and optimization techniques similar to those in the discrete manufacturing industries. The special characteristics of a refinery with respect to continuous flow and mixture of materials result in scheduling problems different to those encountered in discrete manufacturing. Some of the techniques for coupling medium-term planning and short-term scheduling, as well as the modeling difficulties associated with differentiating between shorter- and longer-term problems are presented. The type of tools available and their integration dependent on the organizational structure of the refining company are presented in this chapter.

1 INTRODUCTION

The following discussion uses the concepts of planning and scheduling. Planning is understood to be the optimization of the raw-materials supply, subsequent processing, and marketing of the finished products over one or more *time blocks* of fixed duration. Start date/time and the variable duration of individual activities as opposed to blocked time periods typify the scheduling function.

Planning and controlling functions deal in time blocks for the convenience of the tools used, the data available, and the need to report to management. A very

Optimization in Industry 2. Edited by T. A. Ciriani and R. C. Leachman

typical exercise for the controlling department is the comparison of planned versus actual results. In the refining industry it is typical to deal with yearly, quarterly, and monthly planning periods. The plan covers the period and the actual results need to be accumulated over the events in real time to provide the necessary data for accounting and comparison.

Scheduling in a continuous process such as a refinery is short-term planning. There is, however, a significant difference between a manufacturing process with its batches of discrete materials and a refinery which is working largely with fluid materials with their associated compositions and properties. Refinery units do work with batches (tanks) of different feedstocks to process units and blending. However, a given tank may be both filling and emptying at the same time. Even if it is only filling, the new material flows on top of the previous contents of the tank (called heel). This heel from a previous blend becomes a component for the following blend.

The process units can also be run in different modes resulting in intermediate products with property differences. The blending of final products is seldom based on batch recipes (similar to a manufacturing assembly), but rather uses specifications on the physical properties of the final blend. The net effect is that any scheduling decision to start running, for example, a given unit in a given manner, becomes quickly blurred further downstream as a result of mixing of intermediate products in tanks. Tank composition determines the properties of the material in the tank. Either the composition and the properties of the components in the tank or the properties of the intermediate blend in the tank must be known to characterize the tank contents in its further use.

It is therefore necessary to trace the composition (and hence the properties) as well as the quantity of a given intermediate material. The problem starts at the front end of the refinery. Raw materials—generally crude oil and its light relative, condensate, or its heavier cousin, residue—can and probably do mix in the crude tanks. At the other end of the refinery, the finished products may be lifted from product tanks, and then a new batch of product may be blended on top of the remaining previous blend. The previous blend is thus a component of the next blend and will influence the new blend with its properties and quantity.

Some of the techniques for scheduling discrete manufacturing processes apply to the refinery scheduling problem, but the varying composition of materials in tanks introduces a complexity which has led a number of refining companies to extend the proven tools of mathematical programming into the scheduling area. In general, this will demand the use of mixed integer programming (MIP) to a greater or lesser extent. In fact, MIP may be theoretically the ideal tool to handle the special characteristics of the refinery scheduling problem and developments such as parallel processing of a multi-period MIP problem give hope that this alternative will ultimately prove to be practical.

A single model does not appear to be capable of covering both the medium-term planning and the short-term planning problems. The decomposition of the problem into connected parts seems to be the logical solution. This also permits the use of feasibility search and optimization tools which are more suited to the specific characteristics of the planning or scheduling problem. Irrespective of the tools which are used in the planning and scheduling areas, they can be used much more effectively if they are closely integrated. Networking allows all computerized refinery functions to exchange data and provide mutual support and control. There remains, however, the hurdle of moving from the agglomerated time frame of planning to the start date/time and variable duration of activities which typify the scheduling function.

2 PLANNING

Planning models employing mathematical programming are used for a variety of purposes in the oil and petrochemical industry. The more important applications are:

Investment analysis, where the economics and processing characteristics of new-plant construction or debottlenecking are evaluated against expected market demands/prices for products using a number of key crude oils as markers.

Budgeting, where the margin on the yearly throughput of different types of crude oils is calculated against the background of processing constraints and the expected product requirements. This application may include multiple refineries and the associated distribution logistics.

Seasonal-change analysis in countries with a pronounced summer/winter shift in product specifications. Multiple time periods linked by inventories with detailed depiction of the processing facilities, crude availability both from contracts and from the spot market, and market forecast data are required.

Crude selection, where crude oils, condensates and residues available on contract or in the spot market are evaluated either individually or as additional material against the background of a reference slate.

Crude allocation for multiple-refinery situations where the crude oils have been previously selected and are now allocated to the individual refineries. The combination of crude-selection/crude-allocation modeling techniques will depend on the individual refinery. In some cases there will be only one model with combined crude selection and allocation. In other cases, the initial allocation made at selection time may be readjusted in a shorter-term allocation model.

Partnership models, where the refinery model is used for checking and negotiating the partners' raw-material supply and product-lifting intentions against their capacity share, the refinery's processing capabilities, and overall capacity. A very similar model can be used at month's end to allocate refinery operating costs based on the actual raw-material and product quantities.

Evaluation of processing crude versus purchasing of products and exchange agreements with other refiners.

Operations planning for a period of one week to one month. In this model the raw materials which have been allocated to the refinery and the required product lifting from marketing are input to a detailed refinery model. The purpose of the model is to determine the operating modes and target-blend recipes for the refinery. The sequence of operating-mode changes and blending operations is not considered. The tuned result of this optimization is the starting point for scheduling.

The refinery planning problem is nonlinear due to the pooling of variable volumes (or weights) of intermediate products with properties which vary dependent on the parent or upstream volumes and properties (Fieldhouse, 1993). This mixing or pooling can be considered in two ways:

Implicitly by maintaining the streams with different properties as separate variables and enforcing disposition relationships of the type:

$$V_{ip} / V_{jp} = V_{ip}^{k} / V_{jp}^{k}$$

where V_{ip}, V_{jp} are the total quantities of components i, j respectively in pool p, and V_{ip}^{k}, V_{jp}^{k} are the quantities of the components i and j in pool p which can go to pool disposition k.

Explicitly by blending the component streams to a pool (i.e. blend) stream:

$$P_{P}^{t} = \sum_i (P_{ip}^{t} . V_{ip}) / \sum_i V_{ip}$$

where P_{P}^{t} is the property t of the explicit pool p, and P_{ip}^{t} is the equivalent property of the ith component. If P_{ip}^{t} is a constant, then a simple linear blending model results. However, for many intermediate products in a refinery P_{ip}^{t} is variable dependent on the parent material. This results in a nonlinear model.

The inclusion of certain simple aspects of scheduling, such as minimum-charge selection constraints of the general form:

$$F_{ij} = D_{ij} (M_{ij} + F_{ij}^{v})$$

(where F_{ij} is the ith feed to the jth unit, D_{ij} is the 0–1 decision variable to run feed i at a throughput of at least M_{ij}, F_{ij}^{v} is the additional variable throughput above the given threshold value M_{ij}),

leads to a nonlinear mixed integer problem. The obvious difficulties of solving problems of this complexity encourages decomposition into planning and scheduling models.

3 SCHEDULING

Figure 14.1 depicts a simplified refinery flow. There are four main stages:

Arrival of raw materials and the selection of storage for them. This will perforce involve the pooling of the present contents of the tank with newly arrived material. Only in the case that the raw material arrives via pipeline will the crude oil/condensate/residue sometimes be fed to the primary distillation units without intervening tankage.

Operation of the refinery distillation, desulfurization and conversion units to produce final product component blendstocks. These units take feedstock from tanks and may either send their intermediate products to rundown tanks or feed them directly to further downstream units.

Blending of intermediates to final products to meet specification. Batch blending takes components from rundown tanks, or directly off the processing units and

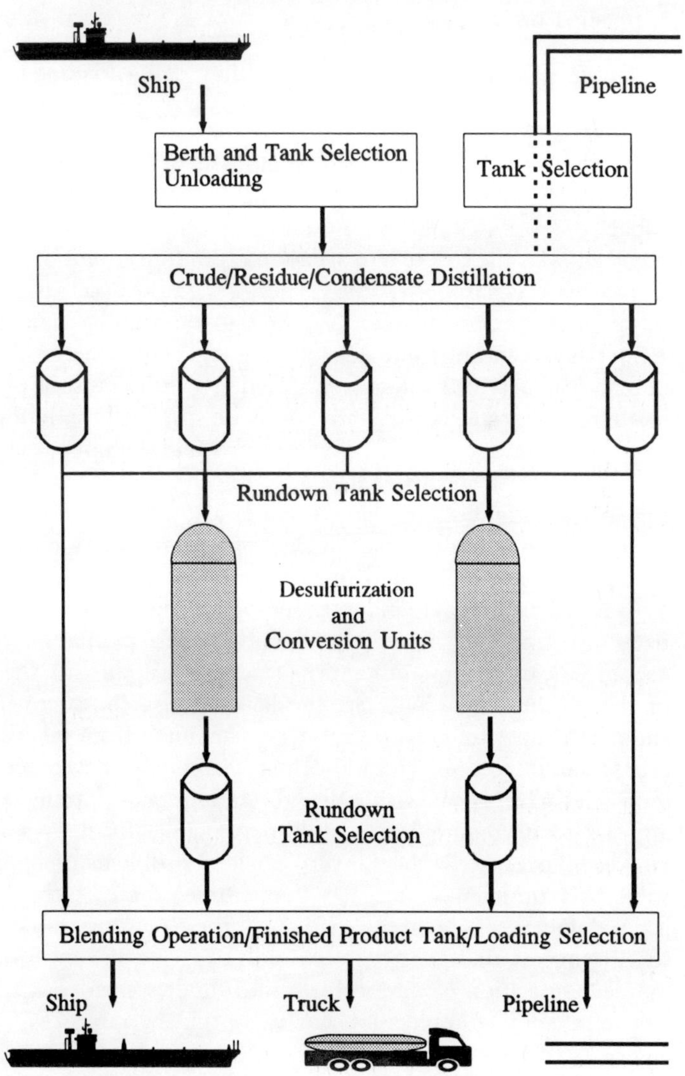

Figure 14.1

blends them into blend tanks. The resultant blend must be certified by the laboratory before it can be moved as final product either into the product tanks or directly to the loading stations. In-line blending is a closely controlled process where the intermediate step of blend (and sometimes component) tanks becomes unnecessary; the product is sent directly to product tanks or even directly to ships which act as a product tank. In either case, the product can only leave the refinery with a Certificate of Quality and must be stored at the refinery in product tanks or other vessels (such as ships) until this certificate is available.

Product lifting refers to the shipment of the finished oil products from the refinery. As noted above, there is generally a time delay between the availability of a finished product and the generation of the required shipping documents.

The constraints of the ship/truck/rail/pipeline loading facilities must be taken into consideration.

Inventory or tankage is a constraint on all of the above activities. Refineries deal with volatile, liquid materials. The initial investment and operating cost of inventory capacity is very significant. Refineries are usually restricted in their scheduling possibilities by the lack of tankage for one or more groups of products, intermediates, or raw materials. Indeed some refineries with the necessary advanced process and oil-movements control systems are designed around a minimum of tankage. Such refineries have a different scheduling goal than refineries designed with a larger tankage capacity; the raw-material feed properties must be precalculated to ensure that the final products are continuously on specification. In this case the scheduling model must be very closely tied to the control and measurement systems of the refinery.

A refinery with adequate tankage, or one in which the tankage is only constraining for a very limited number of materials, faces a problem more typical of manufacturing industry: that of keeping inventory to a reasonable level to ensure minimization of the capital tied up in inventory.

As noted previously, refinery planning models can have many different purposes from investment to operations planning. These planning models make excellent use of linear programming or successive LP due to the continuous nature of the production. However, the lack of discrete intermediate products with constant properties prevents refinery scheduling from taking direct advantage of the scheduling tools and modeling techniques developed for those process industries which deal with either discrete or pure intermediate products. What appears to be required is a multi-period (e.g. 30-day) nonlinear MIP model. However, irrespective of the very obvious problems of finding an algorithm to solve this problem in an acceptable time, there is the question of ease-of-use and understanding of the results. For these reasons, the development of refinery scheduling has alternated between tools to create and solve very large models and the decomposition of the problem into models each with a different objective.

The scheduling application has two primary purposes:

to map planned activities into a sequence of discrete actions of at least a minimum duration while respecting the tankage and other time-related operating constraints of the refinery, to respond to changes in the (monthly) plan and the nominated date/time of receipt/shipments with the rapid generation of a new operating schedule.

A single schedule (which will be subject to roll-forward adjustment) to achieve the targets set by the operations planning model is the objective of the scheduling application.

4 SOURCING AND PIPELINING

Both of these concepts come from the terminology of planning and scheduling in the manufacturing industries. They introduce a brief comparison of optimization tools and their use in refineries as compared to the discrete process and manufacturing industries.

Scheduling tools developed for discrete manufacturing, where raw materials and intermediate products are described fully by their type (i.e. 'pure' products), cannot be used easily for refinery scheduling, where the streams are defined and blended by property on either a volume or weight basis. The production of lubrification oils is an exception; lube oils are usually produced from specific intermediate oils as pure products. They are blended on a recipe basis and subsequently packaged. For this reason lube oil refineries have been among the first to look closely at using discrete manufacturing scheduling tools for their process scheduling.

Sourcing may be described as the medium-to-long-term allocation of manufacturing tasks to individual manufacturing sites to supply markets or other manufacturing sites with their intermediate or final product requirements.

Pipelining refers to short-term throughput planning of individual process lines or distribution facilities. The short-term plan is essentially a schedule giving process or distribution sequences and the timing of changes between these sequences.

In refinery operations planning terms *sourcing* is equivalent to the selection and allocation of the crude oils/condensates/residues available to the refinery. For a large refining company the concept can be extended to the allocation of regional market supply to individual refineries within the company. *Sourcing* is an ideal application of mathematical programming. Over the last 25 years optimization has proven an indispensable tool to the refining industry for this application. The modeling task has been generalized to the point where a number of refinery modeling systems are commercially available which permit definition, generation, solution and reporting of a given refinery model via entry in data tables only. The many specific details associated with the manufacturing industries make this form of generalization more difficult. For specific applications in these industries much effort has gone into developing matrix-generation tools with facilities for setting up project-specific user-interface, data-management, and reporting facilities.

The situation is reversed for *pipelining*. Over the years various industries have tried using time-staged MIP models for the purposes of production scheduling. Depending on the specific industry, these models were partly successful. Some failed due to very long compute times or the inability of the scheduler to comprehend the schedule which had been established in the optimizing 'black box'. Starting in the early 1980s a number of generalized scheduling tools were developed and successfully applied to the discrete manufacturing industries. These modeling systems are interactive, allowing the user to make manual changes to both the basic input data and the schedule and to invoke heuristic probing algorithms to try to improve the schedule. The scheduler can readily comprehend the generated schedule using the graphic displays which include Gantt charts showing graphically the activity sequencing on the production resources over the scheduled period, together with the associated growth/depletion of inventories. The number of probes or iterations is under the control of the scheduler. Knowing when these algorithms will be of value in improving the schedule demands experience both with the heuristics and their applicability to the model. The target used in probing is typically to minimize the sum of

- the penalties for inventory over-/understock,
- the cost of operating processes at different times (e.g. normal, overtime),
- the setup costs associated with process changes,
- the penalties for idle process time.

Using this target, a model can be run to obtain a feasible solution only (the sum of the penalties is zero), or the operating costs of the processes and the switching between activities on those processes can be minimized also to the point where the scheduler is satisfied with the schedule. The concept of a global optimum does not really enter into consideration.

The success with this type of approach in the refinery industry has been limited to date. Refinery processes are by definition continuous and must be run at a certain minimum capacity consumption to be operable at all. They cannot be started up or shut down for the purpose of optimizing a schedule. The single exception is blending. However, product blending in refineries is specification-driven with the additional objective of minimizing give-away on specifications such as gasoline octane number. In other words, the successful discrete-manufacturing scheduling systems contain many features which are of no value to refinery scheduling and are perhaps lacking in certain elemental features such as the ability to track composition in tanks. Nonetheless, the concept of a fast interactive system with easily comprehendible graphic displays, which allow intervention by and transparency to the scheduler is as desirable for refineries as it is for the discrete manufacturing industries.

5 INTEGRATING REFINERY PLANNING AND SCHEDULING

Figure 14.2 shows the linkage between various tasks (in rectangles) and the resultant plan, schedules, and refinery operating data (oval) associated with refinery planning and scheduling. It represents a functional view of integration irrespective of the type of tool used for each of the tasks. The basic principle behind Figure 14.2 is that there is not a single optimization model which can be used to handle both planning and scheduling. However, the use of optimization and, in particular, mathematical programming is foreseen wherever possible.

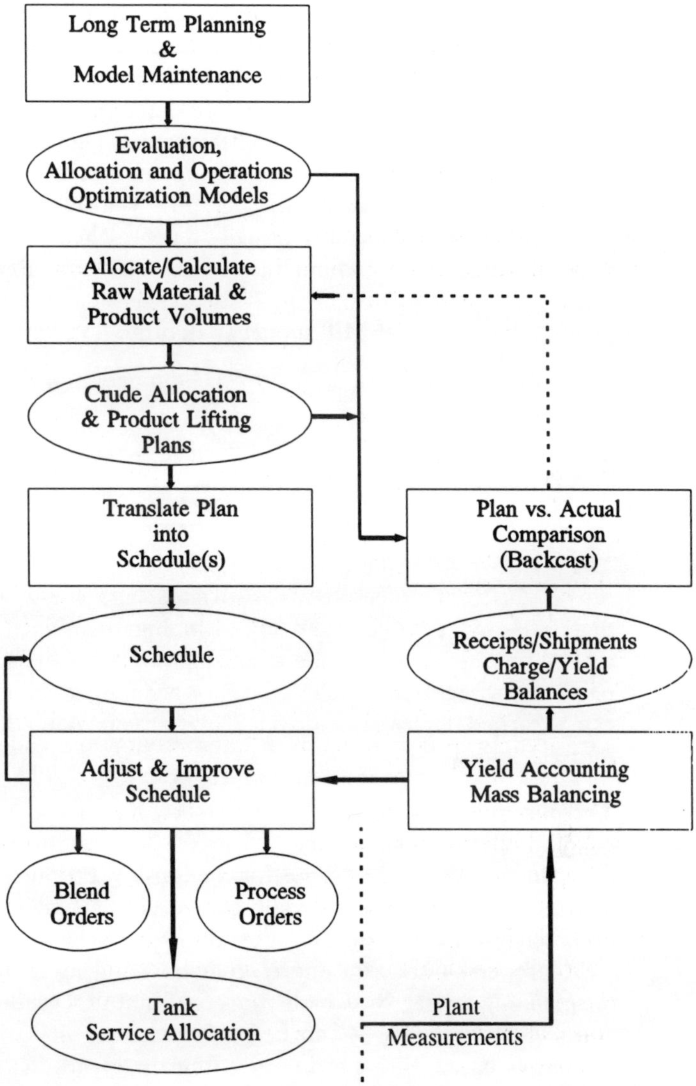

Figure 14.2

The functional integration is essentially cyclical. The elements outside the cycle are the generation of long-term supply/distribution contracts and investments, as well as the generation and maintenance of the refinery-operations planning model. This nonlinear (because pooling is taken into account) model is normally set up as a linear program and the pooling nonlinearities are taken into account using recursion and tracking (Fieldhouse, 1993). The operations planning model is used to optimize the allocation of crude oils/condensates/residues to the possible operating modes of the primary distillation units (see Figure 14.1), to determine the operating characteristics of the desulfurization and conversion units, to calculate target recipes for the product blends, and to establish the feasibility of minimum product lifing as well as any excess production capacity. It will also be used in a slightly different variant to make the backcasting plan versus actual comparison later in the operations planning/scheduling/reporting cycle.

The result of the forecasting optimization is a plan with the allocation of crude oils, downstream process-unit operating modes, product quantities, and the associated target recipes. Once this—typically monthly—plan has been accepted it must be translated into a—typically daily—schedule. At this point two basic modeling options or a combination of both are available:

a multi-period LP or MIP model (Ballintijn, 1993),

an interactive scheduling tool using a combination of heuristic and mathematical programming algorithms.

The operating plan must be translated or 'broadcast' into the scheduling model. This step serves two purposes:

The plan is translated into a starting schedule by taking the production activities on the different process units associated with the volumes (weights) of raw materials and products and apportioning them into the time slots of the scheduling model to provide a starting point for the scheduling task. This is particularly important if an interactive scheduling tool is in use. The apportioning can be carried out by a rule-driven expert system taking tankage limitations partly into account.

The plan provides the yield and property data for the pure or blended crude oils and operating modes selected by the operations-planning optimization. It also provides the target blends for the finished products. These targets may be used as recipe activities. Normally, however, they will be employed in deviation minimization in the blending portion of the scheduling model. By using the yield/property results for mixed crude-oil/condensate/residue operations from the planning model, the simplifying constraint that crude oils must be processed 'pure' (Ballintijn, 1993) can be respected since there is a single set of yield properties based on a previously optimized blend.

It is unlikely that the first schedule calculated using either type of scheduling tool will be satisfactory. Successive fine-tuning by the scheduler will be required. For this reason, and to provide fast response to later rescheduling requests, considerable computing power is required. The concept of a client/server workstation with high floating-point power is proving to be a cost-effective approach to this need. It is also mandatory that the scheduler has the capability to understand quickly the schedule proposed by the tool and to make manual adjustments based on expert knowledge. It is virtually impossible and, at least impractical, to include all the special constraints and vagaries of operating the refinery, the tank farm, and the loading/unloading stations in the scheduling model. For this reason, the concept of an interactive scheduling tool with embedded optimization algorithms appears preferable to a batch LP or MIP model.

The output of the scheduling task can be divided into four broad categories:

receipts giving date/time of vessel unloading or pipeline receipt and tankage to be used,

process orders including start and changeover date/time, feed/rundown tanks and specific operating parameters,

blending orders including component, blend and product tanks, start and finish date/time, and expected blend properties,

shipments giving lifting date/time and tankage or in-line blending facilities to be used.

These orders include the tank service (usage) allocation and form the basis for operations in the process and oil-movements control rooms.

Yield accounting and mass balancing serve an important support role in the integration of planning and scheduling. They have the task of collecting the measured flow and tank gauge data, reconciling the measurement errors and producing aggregated shift, daily, and month-to-date figures for receipts, unit charge/yield and shipment reports as well as balances around required refinery areas and the refinery as a whole. This system must be able to track tank compositions and provide latest tank contents to the scheduling model which will be run usually in a role-forward fashion (and thus requires for each new starting day the latest tank contents) as the month proceeds. The accumulated values for the month provide the actual raw-material intake and product-lifting data for the plan-versus-actual comparison.

An obvious comparison is that between the plan figures as developed in the optimization of the operations planning model and the actual figures as developed by the yield accounting function. An additional level of comparison and checking is available by entering the crude oil/condensate/residue actually run over the month (quarter) as fixed input into the operations planning model. An optimization on this basis will provide a result (often termed *backcast*) with two uses:

A comparison of the backcast with the actual results will provide a check on the accuracy of the model and give an indication of any model maintenance required.

If the accuracy of the model is not in doubt, then reasons must be sought as to why the refinery production did not meet the plan. Of course there can be many reasons why the refinery did not meet the plan, but identifying these reasons has a positive effect on the planning/scheduling operation and serves to tune the models for better-integrated performance.

Each month brings a new cycle through the operations planning model, the starting schedule, the roll-forward schedule, the orders for the refinery, the yield accounting, and the plan-versus-actual comparison. There are many aspects of refinery operation which will affect this cycle, either compressing it into shorter time frames than those considered here or making subcycles within the month necessary. There is a large incentive to handling these aspects and unforeseen situations via an integrated refinery information and control system with a logical refinery database which can be updated essentially in real time via data acquisition and entry transactions. This logical database may have data areas which reside in the individual applications only, or it may draw on a physical relational database.

The concept of a close link between the planning and the controlling functions with satisfactory backcasting results demands an equally close link between the medium term plan and its translation into daily schedules.

6 INTEGRATION VARIANTS

The model structure and algorithms used in refinery planning and scheduling will depend on the type of refinery, its location, and the company policy for its operation. This section looks more closely at one variant to illustrate some of the details of integration.

Figure 14.3 shows a variant in which the blending operations have been identified as demanding an individual blend-optimization model. There are a number of good reasons for singling out blending including government regulations based on environmental issues which are constantly tightening the specifications on refinery products. Two examples are the use of aromatics components in gasolines and the sulfur content of city diesel fuels. Others are:

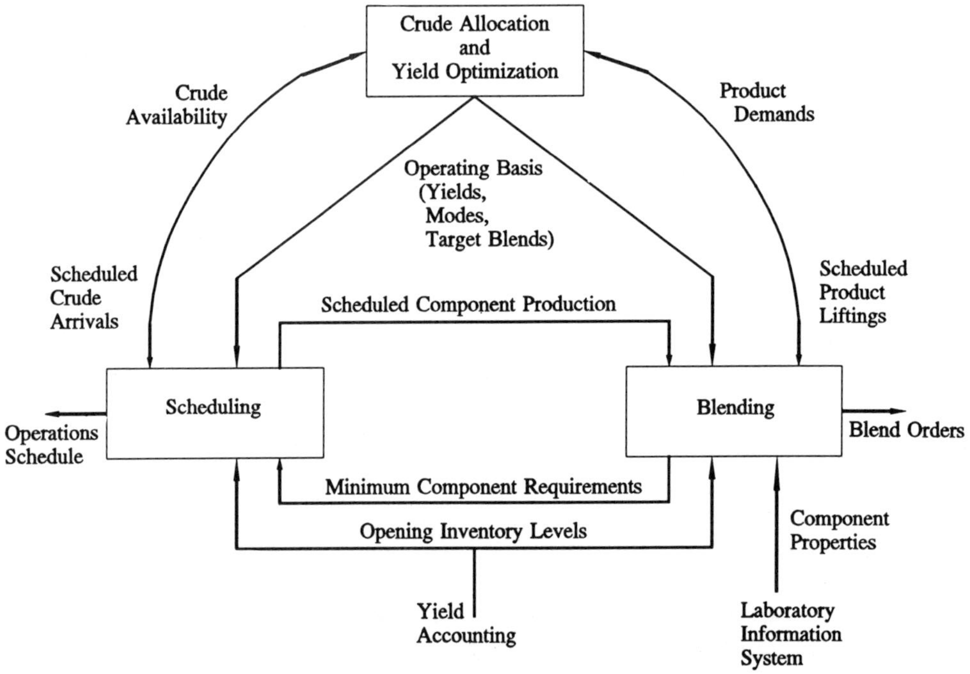

Figure 14.3

One of the most important specifications for gasoline is octane number which certainly does not blend linearly. The monthly planning model uses a linear approximation which is sufficiently accurate for medium-term planning. However, issuing a blend order (from scheduling) for a gasoline demands an accurate recipe which will be translated into setpoints for the flow controllers on the components. One method of handling the nonlinearity is based on laboratory observed values for the interactive effect of components on octane:

$$\mathrm{PO} = \textstyle\sum_i (V_i O_i + \sum_j I_{ij} V_i V_j)$$

where PO is the predicted octane of the blend, V_i, V_j are the volume fractions of the ith and jth components respectively in the blend, O_i is the average octane of the ith component, and I_{ij} is the interaction coefficient between the ith and jth components.

A blend-scheduling procedure must deal with the problem that different grades of product will probably be blended in the same tank over relatively short periods of time. Even if this is not the case, each blend of a product in a tank has its specific set of properties which it passes forward as a remaining volume in the tank to further blending in that tank. In a multi-period model the tank carryover from one time period to the next introduces nonlinearity of the form:

$$P^t_{rp} = \left(\textstyle\sum_i P^t_{ir} V_i + P^{t-1}_{rq} H^{t-1}_q\right) / \left(\textstyle\sum_i V_{ip} + H^{t-1}_q\right)$$

where P^t_{rp} is the property r of product p in time period t, P^t_{ir} is the equivalent property of the ith component, P^{t-1}_{rq} is the property r of the blend q from the previous period, and H^{t-1}_q is the volume of product q remaining in the tank from the previous period.

These and other reasons focus the attention of many refineries on accuracy when blending to meet final-product-blend specifications according to recipes given by the scheduling model. To achieve this goal, blending must be incorporated into the scheduling model with considerable accuracy; recipe blends determined in the planning model are not sufficient.

Two different approaches to this problem use:

multi-period blend-optimization systems with data-integration links to the planning and scheduling models (this is the configuration shown in Figure 14.3);

scheduling system with embedded blend optimization.

The second concept has a great deal of appeal. It seems to cover both problems at once. However, the problem is exceedingly complex and demands algorithms and a user interface which achieve rapid *and* transparent scheduling. The first approach makes use of refinery planning experience with optimization tools. It is, in some ways, a natural extension of the monthly planning model. In both cases maximum use of mathematical programming techniques is made.

The blending-optimization model shown in Figure 14.3 can be developed using commercially available blending-optimization systems. The paradigm is very similar to the refinery modeling system mentioned previously: a multi-period LP formulation with the nonlinearities handled via successive linear programming techniques. This ensures that those in the planning department who are familiar with the refinery-operations planning model are also familiar with the techniques used in the blend optimization. The time parameter is handled by time blocks of a length suitable for the scheduling nature of the model (e.g. 5×1 day $+ 2 \times 1$ week). While the resultant multi-period model does not represent a truly dynamic time scale, the short-term accuracy is well under the control of the scheduler.

Figure 14.3 shows a considerable degree of data interchange, i.e. integration. At the apex of the pyramid lies the refinery-operations optimization model. Essential data inputs are, for example, the nominated crude-oil/condensate/residue availabilities/costs and the product demands/prices. The result of the optimization and its use in setting up the initial schedule have been discussed previously. In the case of the blending model, the product blends determined by the planning model represent targets for the scheduling function with associated component properties. The blending model has, as one of its goals, the minimization of the deviation from the planning-model target recipes. It can use the component properties from the planning model wherever these are required as estimates (i.e. not available as laboratory-measured values for components actually in the tanks). The scheduling model on the other hand can make good use of the process unit yield vectors for different feedstocks to the process units.

In the variant represented by Figure 14.3, no blending takes place in the scheduling model. The scheduling model accepts component requirement data from the blending model and returns the scheduled component production to it. Estimated component quality for components not in tanks (and hence not measurable by the laboratory) are provided by the refinery planning model or come from historical data sources. The scheduling model is an input/output model which takes inventory/tankage into account and seeks to minimize, *under the control of the scheduler*, the effects of tankage overrun/underrun and operating-mode switching while respecting the constraints of raw-material receipts and component requirements. The scheduler can change process/blend run lengths and sequences, and can add or delete activities at specific date/time points or nominate different raw-material arrivals or product lifting. A combination of probing algorithms and manual intervention by the scheduler should permit rapid identification of a viable schedule within the targets set by the planning model.

This integration variant has the advantage that the scheduling and blending tasks are coupled via the component requirements only. However, the prediction of component quality at some future time as the result of a proposed operating plan is left to the refinery-optimization model. If the scheduling model moves outside the space defined by the yield vectors received previously from the operations-planning model, then this model will need to be rerun to provide the necessary new data to the scheduling model.

7 CONCLUSIONS

Refineries the world over produce very similar products from crude oils, condensates and residues which are traded internationally. Refinery processes are limited in number and a specific refinery line-up normally belongs to a generic group. Within a generic group the differences between refineries are their location (e.g. coast, inland), the degree to which tankage is available as an operating buffer, and the operating culture which is imposed by the parent company. The planning problem can and has been packaged. The scheduling problem must respect the specifics of the refinery and its supply/demand to a greater extent and is consequently more difficult to package.

If tankage is generous and supply/demand are relatively stable, then the planning tool can be used in a scheduling role also. Otherwise, scheduling assumes major significance and requires its own set of decision tools. Fast, interactive, graphical systems which provide results *in an easily interpretable form* offer a new dimension over and above the classical use of mathematical programming. Integrating or embedding optimization—especially for the blending area—into the scheduling task is mandatory. Both refinery planning and blend optimization will benefit from advances in algorithms for the solution of the nonlinear MIP problem.

REFERENCES

Ballintijn, K. (1993) Optimization in refinery scheduling: Modeling and solution, in: T. A. Ciriani and R. C. Leachman (Eds.), *Optimization in Industry*, John Wiley & Sons, Chichester.

Fieldhouse, M. (1993) The pooling problem, in: T. A. Ciriani and R. C. Leachman (Eds.), *Optimization in Industry*, John Wiley & Sons, Chichester.

Bonner and Moore Associates GmbH
Bahnhoffstrasse 44/46
65185 Wiesbaden
Germany

15
Algebraic Formulation of Mathematical Programming Models

Tito A. Ciriani and Stefano Gliozzi
IBM Semea, Rome, Italy

ABSTRACT
Modeling concepts introduce the language requirements for analysis and solution of mathematical programming problems. The interaction between model formulation and model instance represents the crucial question of modeling approaches. Model formulation requirements can be satisfied by adopting an algebraic model language and system-integration capability needs efficient data handling. The evolution of modeling systems and present trends are discussed.

1 INTRODUCTION

With a formulation process we intend to analyze a problem into single components and to reassemble them into a model that satisfactorily reproduces the behavior and results of the problem. Mathematical programming (MP) model formulation efforts concentrate on finding the elementary data and activities (with their bounds), on defining the constraint relationships between them, and on selecting the optimization objective.

MP modeling methodologies rely on a database interface, optimization control, and report preparation. Database interface was predominant in the early modeling tools; as reference see B&M (1991), HS (1987), and Day and Williams (1982). Graphics modeling expands the usability of the table-handling approach (Ma Pai-Chun *et al.*, 1989; Chesapeake, 1990). Both these approaches are more suited for (user) database management and operate under an optimization environment.

Because of the growing complexity of real problems (many structures merged in a single model) and the need to interface with several database systems (Tomlin and Welch, 1990; Murphy *et al.*, 1990), algebraic languages have recently been

Optimization in Industry 2. Edited by T. A. Ciriani and R. C. Leachman

developed to translate a model into a code able to interface with data and to prepare the input for an optimizer. Information-system integration requires the batch use of languages.

Linear programming (LP) for real and integer variables adopted the MPS standard input format (IBM, 1988) accepted by most commercial codes. Quadratic programming input format has been introduced by OSL (IBM, 1991). Difficulties arise for nonlinear optimization (Conn *et al.*, 1992) and for heuristic techniques (Pirlot, 1992). Column-generation techniques (Anbil *et al.*, 1992) and network structures (Tomlin, 1983) require a representation of internal data and efficient interface to the optimizer.

In the late 1980s algebraic languages appeared, like GAMS (Brooke *et al.*, 1988), DASH XPRESS (1991) and Maxima MPL (1991). Independence from data has recently been acquired by SML (Geoffrion, 1992) and ModelTalk (Minkoff, 1992) followed by AMPL (Fourer *et al.*, 1993) and EasyModeler (IBM, 1993; De Greef, 1993).

2 MODEL FORMULATION

Abstraction is an important phase to identify the structures of an MP model. Integer programming model structures (constraints) have been defined by Nemhauser *et al.* (1992). We can extend this definition, introducing a structure as a linear expression containing data, variables, and linear operators (see Appendix).

Examples of MP model structures can be found in Shapiro (1984), Mitra (1988), and Williams (1993). Model formulation of real problems in actual size and difficulties are reported in Naylor and Thomas (1984) for finance, Laurence and Zanakis (1984) for manufacturing, Edgar and Himmelblau (1988) for chemicals and Ciriani and Leachman (1993) for manufacturing, transportation, distribution, and petrochemicals.

In these examples, model formulation comes before data considerations. Data tables are used to illustrate the link between the indexed variables and data, with the final aim being to specify a class of constraints or the objective function. The data are seldom reported in entirety since they belong to real applications and usually contain a company's classified information. Williams (1993) lists them for teaching examples. In conclusion the formalization is always part of the descriptive paper, while the data characterize the model development. In addition the early applications were based on a few model classes, mainly to deal with transhipment and blending problems.

Such different roles of formulation and data suggested the implementation of modeling tools where data represent the kernel of the model and constraints are indirectly defined by using the 'block matrix' representation of a selected class of model. B&M GAMMA2000 (1991) and HS OMNI (1987) are well-known examples of block-matrix languages. In such a direction, Johnson (1989) introduces the principles of database modeling, with the flexibility to represent several classes of models.

Limited to education purposes, modeling tools accepting algebraic expressions explicitly written with no summation and indices have been proposed. Modeling feature of LINDO (Schrage, 1991) and MP for PC (Wisniewski and Dacre, 1990) are dedicated to that approach.

More flexibility and more abstraction are required to implement real MP models. Prototyping becomes an important practice of modeling and integer programming problems often require model reformulation to increase the solution performances (Johnson, 1989). When model reuse and documenting become predominant, algebraic formulation is a commonly accepted solution to formulate a model. For details refer to Eppen and Martin (1987) and Temaplan (1992).

Identification of the indexed data and variables is the abstract effort required. Relations (constraint and objective function) are then formulated using summation of indexed objects. Summation is the main operator of the algebraic modeling tools. After the term SIGMA introduced by MAGIC (Day and Williams, 1982), the word SUM is commonly adopted, as in GAMS, XPRESS, MPL, and EasyModeler.

Clauses of constraint and summation are used to limit the number of variables and relations. Extension of the incidence matrix concept is usually used to compact relations and variables, mainly for models that have a close relation with a structured graph. Some examples are given by transhipment, multicommodity, strategic multi-period, and lot-sizing models (Eppen and Martin, 1987). Subset of indices is another option of the clauses as supported by GAMS and EasyModeler.

The modeling tools have spread to the PC (Sharda, 1992). As a result, interpreters are implemented to widen their usage. Interacting with new generation of optimizers the workstation expands towards decision-support systems, usually in client/server local-network operations. The model is compiled into a code able to generate a model matrix from each data set corresponding to the tables and scalars used in the formulation. MGG (SD-Scicon, 1991) introduced a mainframe procedure generating a Fortran code; EasyModeler produces a C code. Figure 15.1 shows the model compilation and run procedure.

3 INSTANCES

When data are closely related to model formulation, the MPS format is no longer an efficient representation of MP models. Network LP problems offer a clear understanding of the need for new standards (Tomlin, 1983). LP recursion techniques (pooling of refinery models) offer another opportunity to extend the data format to revise the original model.

Data format is more and more affecting the model formulation and solution with data-independent modeling as suggested by Murphy *et al.* (1990) and implemented by SML, ModelTalk, and EasyModeler. Model instances, such as set cardinality and data-table size, need not be defined during the model-formulation phase. This architecture separates the model-structure formulation from the data needed to single out a particular model matrix. It paves the way to model reuse and to system integration. Optimization tuning and model documenting are also facilitated.

Database and user subroutines are usually the source of elementary model data. Flat files bridge the instances to generate the model matrix, and the language

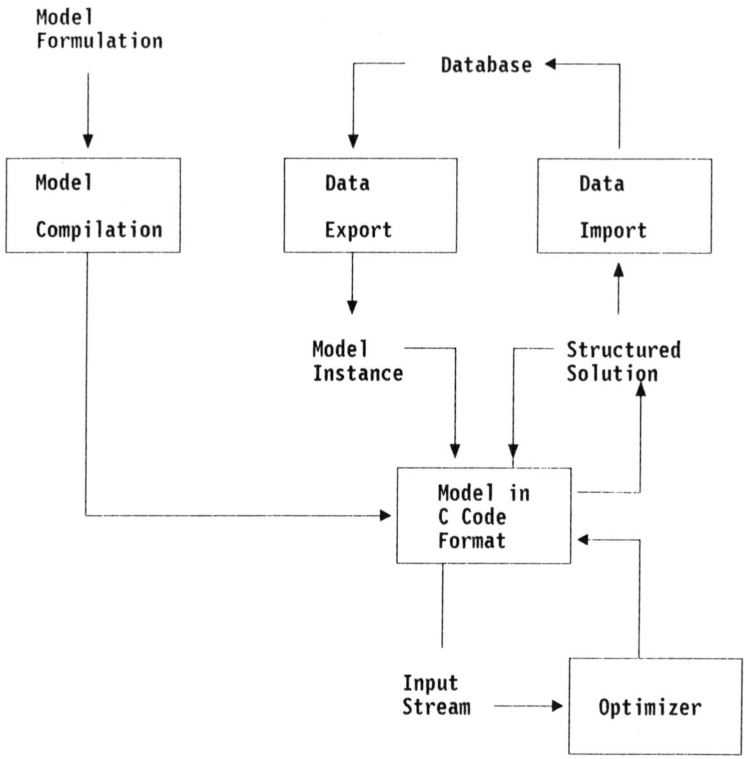

Figure 15.1

includes data manipulation to obtain the matrix coefficients when needed (Internaldata of EasyModeler).

Model languages generate an MPS file to be optimizer-independent but performance considerations and recursion techniques require a close link between external data and the optimizer. For instance, GAMS and LINDO pipe the model matrix into MINOS and OSL, EasyModeler into OSL. GAMS and EasyModeler alternatively generate an MPS file.

Data sparsity is a peculiarity of optimization matrices. Tables (and flat files) may have the same characteristics. Sparse format (index and value) interfaces easily with a database system and increases information-system integration. EasyModeler accepts sparse input format with index specified by numbers or names. In order to hold a reasonable performance with large-model matrices, sparsity should be maintained during the input-generation phase. For instance, EasyModeler compresses internal data into a supersparse format.

Tables, arrays and scalars are the data structures used in the model formulation. Solution structures (indexed variables) represent the expected output of a modeler. But the optimizer provides a solution report with only two sections: columns (structural variables) and rows (logical variables). For a clear understanding of the optimization results, the modeler should be able to re-collect those variables in the structures that have been defined by model formulation. In addition, the interest is usually concentrated on nonzero values. Then the solution structures are preferred in a sparse format, like the input data structures, mainly for database integration purposes. In particular, EasyModeler speeds up optimization recursion with direct OSL dataspace access: it combines supersparse internal data handling with a sparse external data interface.

In conclusion, model instances change with the application of a specific model class. They are transformed into an input stream which can be independent from the optimizer (like an MPS file) or piped into a selected optimization tool, like DSPACE of OSL (for instance, XPRESS and EasyModeler).

Figure 15.1 shows the modeler's data flow.

4 OPTIMIZATION PROCESS

The model (identified by the generated code), the instances, the optimizer, and the structured solution are the objects handled by an application. A strategy combines these objects to solve a particular class of problems (Figure 15.1).

The straightforward strategy with no optimizer dependency can be represented by Figure 15.2. Recursive LP can be achieved by the strategy of Figure 15.3. Figure 15.4 shows a strategy to iterate over several models. Such strategies can be implemented using algebraic languages. EasyModeler supports them in the IBM RISC S/6000 batch environment.

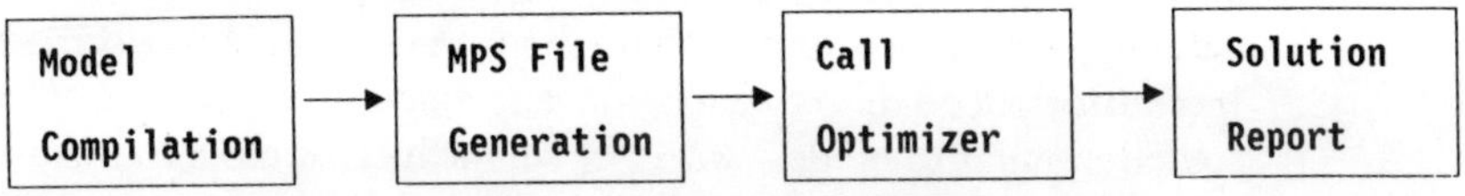

Figure 15.2

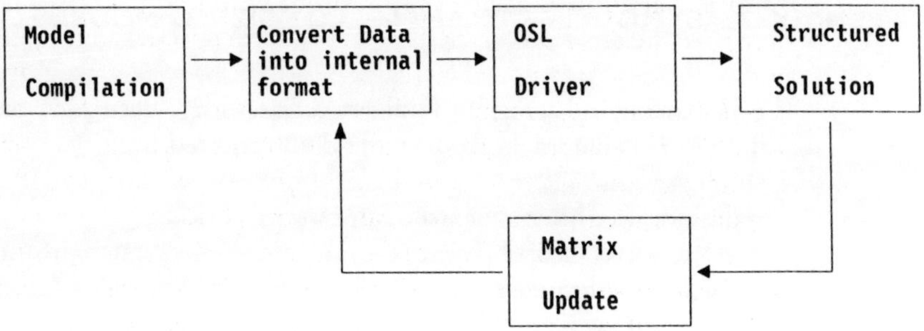

Figure 15.3

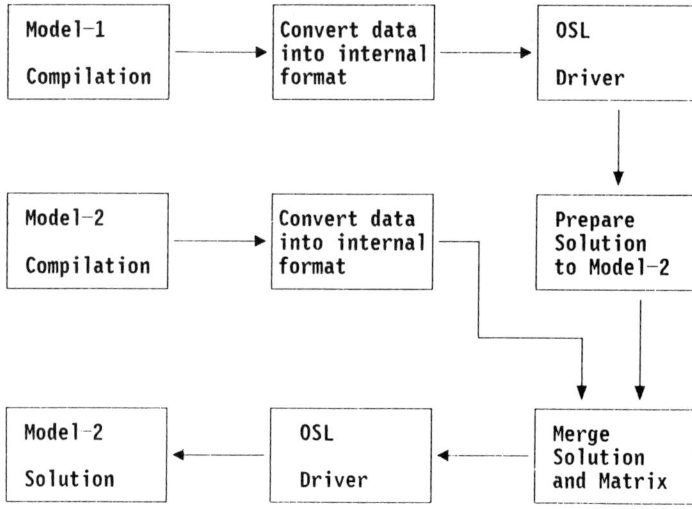

Figure 15.4

5 FINAL REMARKS

Large and complex problems require more flexible and efficient MP modeling tools operating on different platforms. Application strategies are easier implemented by modeling tools if they support a batch integration in the company's information system. Modeling systems support algebraic language, database interface, structures object handling and integration with optimization subroutines. Cooperative and parallel processing is encouraging the extension of object-oriented strategies. The specialization of the optimization approaches requires languages adapted to a predefined class of models, but open to unexpected model structures. The improving performances of the optimizers recommend modeling languages able to combine information-system integration, model-formulation, and report-analysis capabilities.

APPENDIX EASYMODELER EXAMPLE: LOT-SIZING MODEL FORMULATION

The lot-sizing model using Boolean variables has been proposed by Eppen and Kipp (1987). For each product the data required are:

- the projected demand over subsequent periods,
- the inventory cost,
- the production cost,
- the setup costs,
- the amount of production capacity used to produce one product unit,
- the production capacity available in each period.

The optimal solution gives the production plan which meets the projected demand at the minimum cost.

The EasyModeler formulation follows.

```
Lot Sizing GENERIC
  |*            A Lot Sizing Disaggregated model              *|
SET

  i     Part_number;            |*Different Products          *|
  p,d,q,v Time_period;          |*Planning Horizon            *|

ENDSET
DATA

  Inventory_u_cost[i];          |*Cost of keeping PN i as     *|
                                |*  Inventory for one period  *|
  Set_up_cost[i];               |*Fixed Cost of Starting PN i *|
                                |*  production                *|
  Demand[i,p];                  |*Projected demand of PN i in *|
                                |*  period d                  *|
  Capacity[p];                  |*Production Capacity in period p*|
                                |*                            *|
  Unit_capa_use[i];             |*Marginal use of Production  *|
                                |*  Capacity for PN i         *|
ENDDATA
INTERNALDATA
  DemandByCap[i,p] = Demand[i,p]*Unit_capa_use[i];
  InvCost[i,p] = Inventory_u_cost[i]*Demand[i,p];
ENDINTERNALDATA

VARIABLE

  Disagg_product[i,p,d];     |*Amount of PN i produced in period*|
                             |*  p to satisfy demand in period d *|
  Decision[i,p] LOGIC;       |*Logic Variable: Produce (=1) or   *|
                             |*  not (=0) the PN i in period p   *|
ENDVARIABLE
CONSTRAINT

  Limited_Capa (FORALL p)
  SUM(DemandByCap[i,v]*Disagg_product[i,p,d])
      (OVERALL i; FOR d>=p; FOR v FROM p TO d)
  <= Capacity[p];

  Balance0 (FORALL i;FOR p=0)
  SUM(Disagg_product[i,p,d]) (FOR d>=p)=1;

  Balance (FORALL i; FOR p>=1)
  SUM(Disagg_product[i,p,d]) (FOR d>=p)-
  SUM(Disagg_product[i,d,p-1]) (FOR d<p)=0;

  Integer_link0 (FORALL i,p;FOR Demand[i,p]=0)
  SUM(Disagg_product[i,p,d]) (FOR d>p)<=Decision[i,p];

  Integer_link00 (FORALL i,p; FOR Demand[i,p]<>0)
  SUM(Disagg_product[i,p,d]) (FOR d>=p)<=Decision[i,p];

ENDCONSTRAINT
OBJFUNCTION
                             |*Minimize the overall production   *|
                             |*cost over all the planned horizon, *|
                             |*i.e. inventory cost and fixed costs *|

MINIMIZE Cost
    SUM(Disagg_product[i,v,q]*InvCost[i,d])
      (OVERALL i,v; FOR q>v ;FOR d FROM (v+1) TO q ; FOR p FROM (v+1) TO d)
    + SUM(Decision[i,p]*Set_up_cost[i]) (OVERALL i,p);
ENDOBJFUNCTION
```

REFERENCES

Anbil, R., Tanga, R., and Johnson E. (1992) A global approach to crew-pairing optimization. *IBM Systems Journal*, **31**, 71–78.

B&M (1991) *GAMMA2000 User Manual*, Bonner & Moore, Houston, Tex.

Brooke, A., Kendrick, D., and Meeraus A. (1988) *General Algebraic Modeling System: A User Guide*, The Scientific Press, Redwood City, Calif.

Chesapeake (1990) *MIMI/LP User Manual*, Decision Sciences Inc., New Providence, NJ.

Ciriani, T. A. and Leachman, R. C. (Eds.) (1993) *Optimization in Industry*, John Wiley & Sons, Chichester.

Conn, A. R., Gould, N. I. M. and Toint, PhL. (1992) *LANCELOT: A Fortran Package for Large-Scale Nonlinear Optimization*, Springer, Berlin.

DASH (1991) *EXPRESS-MP Reference Manual*, DASH Associates Ltd, Blisworth, UK.

Day, R. E. and Williams, H. P. (1982) *MAGIC: The Design and Use of an Interactive Modelling Language for Mathematical Programming*, Technical Report 2/82, University of Edinburgh.

De Greef, G. (1993) *Analyse Comparative de Deux Logiciels d'Aide a la Modelisation en Programmation Linéaire*, Thesis presented at the Université Catholique de Louvain, Belgium.

Edgar, T. F., and Himmelblau, D. M. (1988) *Optimization of Chemical Processes*, McGraw-Hill, New York.

Eppen, G. D., and Martin, R. K. (1987) Solving multi-item capacitated lot-sizing using variable redefinition, *Operations Research*, **35** (6), 832–848.

Fourer, R., Gay, D. M., and Kernighan, B. W. (1993) *AMPL A Modeling Language for Mathematical Programming*, The Scientific Press, San Francisco.

Geoffrion, A. M. (1992) The SML language for structured modeling. *Operations Research*, **40** (1), 38–75.

HS (1987) *OMNI Linear Programming System, User Reference Manual*, Haverly Systems, Denville, NJ.

IBM (1988) *Mathematical Programming System Extended/370 Version 2, Program Reference Manual*, SH19-6553, IBM Corporation.

IBM (1991) *Optimization Subroutine Library Release 2, Guide and Reference*, SC23-0519, IBM Corporation.

IBM (1993) *AIX Easy Modeler/6000, User Guide*, SB13-5249, IBM Corporation.

Johnson, E. L. (1989) Modeling and strong linear programs for mixed integer programming, in: S. W. Wallace (Ed.), *Algorithms and Model Formulation in Mathematical Programming*, Springer, Berlin.

Laurence, K. D., and Zanakis, S. H. (1984) *Production Planning and Scheduling: Mathematical Programming Applications*, Ind. Engl. & Management Press, Atlanta, Ga.

Ma Pai-Chun, Murphy, F. H., and Stohr, E. A. (1989) A graphic interface for linear programming, *Communications of the ACM*, **32** (8), 996–1012.

Maxima (1991) *MPL Modelling System*, Maxima Software Ltd., Arlington, Va.

Minkoff, A. S. (1992) A systematic approach to OSL application programming, *IBM Systems Journal*, **31** (1), 71–78.

Mitra, G. (1988) *Mathematical Models for Decision Support*, Springer, Berlin.

Murphy, F. H., Stohr, E. A., and Asthana, A. (1990) *Representation Schemes for Mathematical Programming Models*, Technical Report, School of Business, Temple University, Philadelphia and Graduate School of Business Administration, NY University.

Naylor, T. H., and Thomas, C. (1984) *Optimization Models for Strategic Planning*, North-Holland, Amsterdam.

Nemhauser, G. L., Savelsbergh, M. W. H., and Sigismondi, G. C. (1992) Constraint classification for mixed integer programming formulation, *Committee on Algorithms Bulletin*, MP Society, No. 20, 8–12.

Pirlot, M. (1992) General local search heuristics in combinatorial optimization, *Belgian Journal of Operations Research, Statistics and Computer Science*, **32** (1–2), 7–67.

Schrage, L. (1991) *LINDO An Optimization Modeling System*, The Scientific Press, Chicago.

SD-Scicon (1991). *Model Generator Generator, User Guide*, Milton Keynes, UK.

Shapiro, R. D. (1984) *Optimization Models for Planning and Allocations*, John Wiley & Sons, Chichester.

Sharda, R. (1992) Linear programming for personal computers: 1992 survey, *OR/MS Today*, June.

Temaplan (1992) *Bewertung von Matrixgeneratoren fuer die Optimierung von Schwellketten*. Technical Report, Vienna.

Tomlin, J. A. (1983) *On the Specifications of Network Models in Mathematical Programming Systems*, Presented at SHARE 61, New York.

Tomlin, J. A., and Welch, J. S. (1990) *Mathematical Programming Systems*, Technical Report, IBM, Almaden Research Center, San José and Silver Spring, CA.

Williams, H. P. (1993) *Model Building in Mathematical Programming*, John Wiley & Sons, Chichester.

Wisniewski, M., and Dacre, T. (1990) *Mathematical Programming*. McGraw-Hill, New York.

IBM Semea
Scientific and Technical Systems and Solutions
Viale Oceano Pacifico 173
00144 Rome
Italy

Index

93,